The Geological Society of America
Life History of a Learned Society

Richard Alexander Fullerton Penrose, Jr., 1863–1931
Geological Society of America President 1930
Generous contributor to the welfare of the science
and of the Society, in perpetuity, 1931

The Geological Society of America, Inc.
Memoir 155

The Geological Society of America Life History of a Learned Society

by
Edwin B. Eckel
with a Foreword by
Robert F. Legget

Published by
The Geological Society of America, Inc.
P.O. Box 9140, 3300 Penrose Place
Boulder, Colorado 80301

Printed in U.S.A.

Library of Congress Cataloging in Publication Data

Eckel, Edwin Butt, 1906–
    The Geological Society of America

    (Memoir ; 155)
    Bibliography: p.
    1. Geological Society of America—History. I. Title.
II. Series: Memoir (Geological Society of America) ;
155.
QE1.E148   1982   550'.6'073   82-15412
ISBN 0-8137-1155-X

To the many devoted women of the GSA staff . . .
only a few appear in these pages, but all have contributed
much to the health and continuity of the Society

Contents

Foreword

"People will not look forward to posterity," wrote Edmund Burke, "who do not look backward to their ancestors." The Geological Society of America, as it approaches its centenary, will assuredly be looking forward, if not perhaps to posterity, most certainly to the challenge of its second century. This volume enables its members to look backward, to examine its foundations, to see how the Society has grown in stature and influence, to see how greatly the Society and so the science of geology have benefited by the munificence of R.A.F. Penrose, Jr. It is a fascinating record.

"The purpose of the Society is the promotion of the science of geology." GSA consists of its members, its history, its traditions. Of itself the Society can do nothing to aid geological progress, but it serves the science through the assistance it gives to its members. First thoughts turn naturally to research grants and publications. The record of GSA in these fields has been notable indeed. The long row of bound volumes of the GSA *Bulletin,* to be seen on so many a library shelf around the world, provides an invaluable corpus of knowledge. Many a younger geologist has been helped along his way at the start of his career by a grant received from the Society.

Important as these functions of the Society are, they are not unique. There are other geological publications, and there are other research-granting bodies. What is unique, however, are the regular meetings of the Society—the annual meetings now dating back almost a century, the highly successful but smaller regional meetings held since 1899, and the specialized Penrose Conferences more recently inaugurated. All too many take these meetings for granted without giving thought to the long background of steadily growing experience that makes them such valuable gatherings, or to the tremendous amount of voluntary work done by local committees, aided by the support of the Society's staff, that ensures their successful operation.

These are pleasant social gatherings but more, much more than this, as the record in this volume makes clear. Inevitably and naturally there are complaints—that meetings are too big, that too many papers are accepted for presentation. But members of the Society attend in ever-increasing numbers, testimony to the value of the face-to-face associations—with friends, with coworkers, even with opponents in academic argument—that such meetings alone make possible. In some ways, provision of the facilities for these personal contacts is probably the Society's greatest contribution to the advance of the science. Some warrant for this view of the importance of GSA meetings is given by the loyal attendance of Past Presidents, no less than 13 of the 20 then living having attended one of their recent early morning annual meeting breakfasts.

The fact that the smooth running of GSA meetings is taken for granted, as well as the provision of all other services of the Society, is well shown by the almost unbelievably small attendance at the Society's annual business meetings, rarely more than a dozen members in addition to officers and members of the Council. In one way this is a good thing in that it indicates that the membership is satisfied with the way the Society is operating, but it is not altogether a healthy sign since it is at this meeting rather than in back rooms and along corridors that the objections to the electoral system of GSA should be voiced, the complaints about too many papers at meetings should be discussed, and the strong feelings against extending the annual meeting beyond the traditional three days should be aired.

This volume includes a lucid explanation of how the electoral system works. The objectors to the system are probably few in number. I wish to record my personal conviction that the system is indeed democratic. The council selects the nominees by secret ballot. One need only compare the Society's system and the results it has given down through the years with the sad and lingering animosities created in other societies that persist in holding an open election for their highest office to see how well served we are in GSA.

I can think of no major Society policy that is not at least touched upon in this volume, and many of them clearly and dispassionately explained with never a hint of the author's own opinions until his useful, interesting, and provocative concluding chapter. The record he has prepared is the more valuable since the *Proceedings* of the Society have been discontinued as a separate publication. This, in my view, a regrettable, even if perhaps inevitable, step as we all who have need to trace any detail of the Society's history will discover, despite the broad coverage of this volume.

The great contribution made by Dr. Penrose is most interestingly demonstrated in the following pages, his gift to GSA still the envy of many other societies. It did indeed mark a watershed in the history of the Society after almost 50 years of solid progress. The record shows also, although not so dramatically, that a second watershed of different but comparable significance was passed in 1962, by coincidence almost exactly at the seventy-fifth milepost in the Society's history (if metaphors may be mixed for once). President King Hubbert in that year persuaded A. Rodger Denison, a former member of Council, to preside over a small ad hoc committee charged with a study of the Society's administration as reflected in its Bylaws. This followed the retirement of Henry Aldrich, whom so many thought of as "the perfect Secretary," as I did also with affection, although later I had to admit that he had allowed the old Bylaws to get a bit ragged.

The shock of the death of Dr. Denison and his wife on the first leg of a joyful preretirement overseas tour will never be forgotten by those of us who had the gift of his friendship. But he had laid the foundation of the work with which he had been charged, especially the major decision to discard completely the older Bylaws and develop a basic Constitution and a new set of Bylaws suitable to the challenges then facing the Society. Roger's colleagues carried on and completed the task, after wide consultation with the membership, their recommendations being accepted by Council and eventually adopted by the Society in the form still in use today with only minor changes.

Reasons for the drastic changes then suggested were given to Councils of the time and later to the 1963 Annual Meeting held in New York (one well attended for a change). Apparently no written record was then prepared, as it should have been. As a small supplement, therefore, to Chapter 6 of this volume, may it now be recorded that the basic premise underlying all the changes then suggested was the placing of full control of all Society business firmly in the hands of a democratically elected Council. It was for this reason that the Committee on Policy and Administration was dropped, since it had become almost a second Council. Correspondingly, the role of the Executive Committee was clearly delineated as that of a small group to deal only with emergency situations and the few necessarily private matters with which all societies have to deal. Every effort was made to ensure that nothing like a "one-man rule" or "the officers running the Society" could ever develop in the future.

It was realized that this would mean an increase in the responsibilities of future Councils, even though they would be aided by the more detailed work of specially appointed committees. The possibility of the Council needing three meetings each year instead of two, as is the practice in comparable societies, was discussed. To protect the special position of all future incumbents of the office of "chief-of-staff" to the Society from any recurrence of difficulties of the past, his title was changed to the honored one of Executive Secretary. The inference was that, as such, he would be the legal "Secretary to the Council"; this should have been stated. And, as is invariable comparable practice, he was not to be a member of the elected Council. A special effort was successfully made regarding one detail—to eliminate for all time the word "proxy" (to which some members had objected strongly)—with the skilled and delightful assistance of Rollin Browne, then the Society's legal counsel.

While all these proposals were being hammered out, the operations of the Society had to be kept running smoothly: committees had to be serviced, including the recalcitrant ad hoc Committee for Revision of Bylaws; skilled editorial work had to continue and the regular issues of the *Bulletin* produced; new accommodations had to be found and eventually the entire headquarters operation moved from Columbia University to the new building on East 46th Street, New York. All this represented enough work for two Executive Secretaries, and the Society did not then have even one. But they did have Agnes Creagh.

This charming lady, now in happy retirement after a second career, will be but an honored name to younger members of the Society. Some older members will remember her skill as one of the best editors with whom they had ever worked. She was the Managing Editor of the Society when called upon without warning to assume in addition all the duties of the Executive Secretary. In the pages that follow will be found a brief, factual recital of what she did for the Society in those critical years of 1962 to 1964. The full story of her unique service can never be told. Even at the time there were only a few members who realized how Miss Creagh with her few associates quite literally saved the Society from disaster. Her service was no less than this.

So to today. Having finally broken away from New York City, after years of discussion, the Society is now well established in Boulder, Colorado, in its own dignified, efficient, and economical building. The only disadvantage of its headquarters is that its location, desirable as it is from so many points of view, prevents most members from seeing and visiting "their building." Beset with financial problems arising from current inflation, as are all voluntary societies, the Society is tackling these with vigor and with success. The now-enhanced Penrose bequest is still intact, and the income alone is being used to further the work of the Society and so is in service to the science of geology. It would surely be a good thing if the Society's accounts always showed the Penrose fund quite separately from its normal assets and income so that all younger members could see how they are the beneficiaries of this gift of half a century ago.

Served by a loyal and efficient staff, the Society is in good shape to meet the challenges of its second century. What this new century will bring for the Society, who can tell? But as they go forward, the members of the future and especially all future office holders will have in this volume an invaluable aid to their thinking. There must be very few scientific societies that have had the great good fortune to find a member so well qualified as Ed Eckel and, at the same time, one so willing to devote his skill and energy to the preparation of so lively and accurate an account of their "Life History."

If I may conclude this Foreword (forced out of me, but so courteously) on a more personal note, I have read every word of this volume with lively interest, with real enjoyment, and with close attention and appreciation of its accurate and discreet coverage. Naturally, I do not agree with everything in the final chapter—who will? But I find this fascinating and provocative, clearly a stimulating challenge to those who will carry the torch into the second century.

My personal recollections of the Society are no match for those of the author; mine go back only to 1947 when, by happy chance, the annual meeting was held in Ottawa, then my new home. Dr. Charles P. Berkey, then in his prime, was my sponsor in Society affairs, and Dr. Henry R. Aldrich was a most friendly and helpful guide to the neophyte. It was in this same year that the Engineering Geology Division of the Society had its first meeting, under Dr. Berkey's stimulus; it is now the largest of the Society's Divisions, the only one with its own series of publications.

Dr. Berkey was one of the towering figures of the first half century of the Society, serving as Secretary from 1923 to 1940 and then as President in 1941. It would be interesting to know how many members of the Society today realize that the subject of the presidential address of this great man was "The Geologist in Public Works"—and this almost 40 years ago. Well worth reading today, the address was percipient in many ways—"Perhaps the salvation of the civilization that we know depends more on these new sources of electric power than most of us are prepared to believe" being but one of his farsighted comments.

Dr. Berkey would have been encouraged, I think, by the results of the survey of the major professional interests of the members of "his Society." Recently more than 40% indicated some branch of applied geology as their principal interest, and 12% have as their major concern "Geology and Public Works." Here may be yet another guidepost for the future.

Ottawa, Canada
28 February 1980

ROBERT F. LEGGET

Preface

My ties to the Society, though tenuous for the most part, go back for nearly 70 of GSA's 92-year history. As the son of a Fellow, from earliest childhood I heard casual mention in family conversations of GSA and of many of its personalities. Thus, long before I discovered geology for myself, I knew that there was such a thing as the Geological Society of America, and I knew by sight or hearsay some of the strengths and foibles of many of its leaders. Much later, in 1939, I was honored by election to Fellowship. I played the usual passive part of the younger set, attended many of the annual meetings, and voted the straight ticket.

As Councilor for 1961–1963, I contributed little directly but did develop a deep interest in the Society's management. At the close of my term I volunteered to codify the welter of rules and customs that burdened and mystified successive Councils, a codification that was an essential corollary to the new Constitution and Bylaws adopted in 1963. The resultant document, *Council Rules, Policies, and Procedures, 1964,* was a minor but important building stone in the Society's management. A few years later, in 1970 and again in 1974, I had the pleasure of helping to produce updated versions of the Rules.

Late in 1967 as the Society was moving to Boulder, President Konrad B. Krauskopf and Executive Secretary Raymond C. Becker approached me with an offer of the Editorship. This offer came out of the blue; I had not aspired to it and, other than a few private remarks as to the requirements of the Editor's job, had shown no interest in it. But the offer came when my U.S. Geological Survey work was at a lull, so I sought early retirement, took the plunge, and started a new life with GSA on February 1, 1968. The initiation was painful. Ray Becker was still as-

sembling and training an entirely new staff, and worse yet, the backlog of manuscripts awaiting any kind of action was mountainous in that Ray had neglected them for months as he tried to cope with the triple job of Executive Secretary, Editor, and prime mover of headquarters from New York to Boulder.

Not only did I learn the job as Editor, thanks to the help of innumerable capable friends among the staff, the authors, and the Publications Committee, but I also attained a good working knowledge of Society management from Ray Becker, the Council, and many others. As a result, when Becker suddenly announced his retirement at the Atlantic City Annual Meeting in 1969, less than two years after I had come aboard, the Executive Committee took the easy way out and promptly asked me to become Acting Secretary while continuing my duties as Editor. Following the example of several predecessors who had been asked from time to time to assume dual responsibilities to avoid or postpone recruiting problems, I accepted.

After a few months' service as Editor and Acting Executive Secretary, I was relieved of the editorship and assumed full-time responsibilities as Executive Secretary in May 1970. On August 30, 1974, I took my second retirement. As the retirement date approached, I offered to try to compile a history of the Society, with emphasis on the later, affluent, post–Penrose bequest years with which I was most familiar. The Executive Committee and the Council agreed, and this book is the result.

Many passages may seem overcritical of individuals or groups and of their actions and decisions. Despite efforts to be objective, my own feelings have unquestionably crept in. I believe, however, that any historian, given the same set of facts available to me, would come to interpretations closely similar to mine.

As I reread the manuscript, I am struck by a feeling that I may not have given sufficient space to the Society's Presidents, other elected officers, and Councilors. True, I have tended to emphasize the predominant roles of the worker bees—Secretaries (Executive Directors), Treasurers, Editors, and particularly their staffs, paid and unpaid, elected or appointed—in molding the Geological Society of America. These people have carried out the policies laid down by successive Councils. In doing so, they have played significant parts in making GSA what it is today and in setting its course for what it will be tomorrow. But this is not to say that I lack respect for the Society's elected leaders. Since I joined GSA, I have seen more than 40 of its 90 Presidents come and go; I served under three of them as a Councilor and under seven as an employee. Some of these I count among my best friends (without the friendly support of several of them I would never have tackled this book). I admire all of them and all of their distinguished predecessors not only for outstanding accomplishments in geology but also for their deep devotion to the sound guidance of the Society while they held office. The elected leaders have all had their share of acclaim in the scientific world. Staff members, on the other hand, have received relatively little notice for the essential parts they have always played in helping the Society maintain its operations and attain its objectives.

Throughout this volume I have tried to treat each facet of the Society's history objectively and to trace its development, although in a very general way, from its inception to the present. In doing so I have in many places tangled my tenses. Specifically, the present tense appears far more frequently than one would expect in a history, which is by definition supposed to be a story of the past. There is nothing wrong with bringing a history up to the present day, except for the author's difficulty in determining a cutoff point. The reader must remember that use of the present tense does not mean that the Society became static when this book went to press. The Society will go on, and changes in its appearance, makeup, and activities will continue to take place, probably more rapidly in the ensuing decades than they have in the past. The only purpose of history, other than to provide a record of the past, is the hope of providing guides to the future, some insight as to what that future may be like, and above all to help prevent repetition of errors that may have been made in the past, some of them more than once.

Without the warm support and patient assistance of many friends this work could not have been completed. Several former long-time GSA employees—Miriam F. Howells, Helen R. Fairbanks, Agnes Creagh, and Marie Siegrist in particular—dug deeply into their memories and their memorabilia to provide facts, characterizations, and color that would have escaped me completely in the written record. They also provided forthright criticisms of some of my early drafts and helped keep me within the bounds of propriety. Similarly, two former Treasurers, J. Edward Hoffmeister and August Goldstein, Jr., kindly read my material on financial management to make certain I had caught the essentials of a complex and touchy segment of the Society's history.

Nearly every present and recent staff member has cheerfully helped me in this compilation in one way or another—in file searching, document copying, and other chores—but above all in interest and in informal but helpful criticism. Among all these, June R. Forstrom, Dorothy M. Palmer, Linda L. Reidler, Fred S. Handy, Lee Swift, and Joan Heckman stand out. My special thanks go to Dorothy Merrifield and Mary Alice Graff for the final editing, to Alison Conn Richards for turning faded photographs into crisp drawings, to Marianne Faber for preparing many of the portrait photographs, and to Claire Zimmerman for patiently typing the growing manuscript time after time. Lee Gladish not only prepared many of the photographs but prepared the book for the printer with sympathy, understanding, and great skill.

Past President Robert F. Legget honored me by preparing the Foreword to this book. Both he and Claude Albritton kindly read the entire manuscript with patience and understanding. They corrected many errors in style and presentation. Far more important, they wisely counseled me on matters of fact and judgment.

I must also thank successive Executive Committees and Councils who encouraged me in an effort that turned out, on the whole, to be personally rewarding. Finally, many Society members have volunteered discussions about GSA's present and future health, welfare, and problems. Together they have helped me keep perspective and to tread the narrow path between adulation of a great institution that I revere and perhaps too-harsh criticism of individuals or groups who have made mistakes and who may do so again.

EDWIN B. ECKEL

Part I

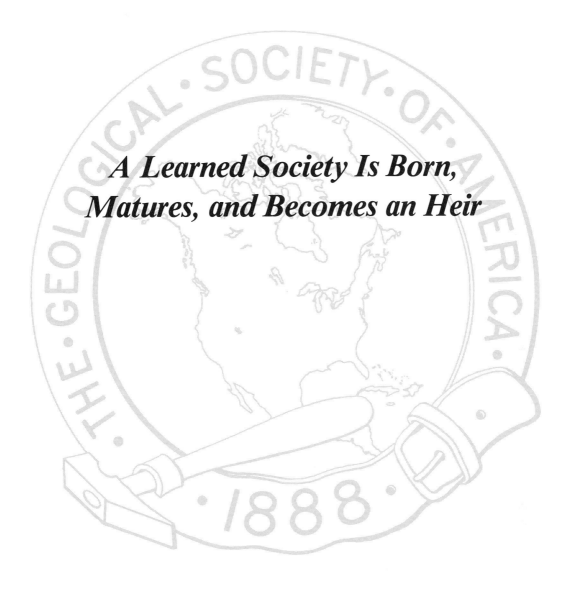

A Learned Society Is Born, Matures, and Becomes an Heir

1

Introduction

On July 31, 1931, a wealthy 67-year-old bachelor died, alone, in his hotel suite in Philadelphia. He was Richard Alexander Fullerton Penrose, Jr. At the moment of his death, he provided for the remolding of the Geological Society of America almost in his own image. Supported heavily by his generous endowment and guided by many of his own principles, the Society has gradually come to resemble him as much as a corporate body can resemble an individual human. Large in stature, both in body and mind; generous, but thoughtfully discriminating in his generosities; dedicated to the advancement of his chosen science and its scientists, though not a heavy contributor to that advancement himself; courteous to all; loyal to friends and relatives; modest and self-effacing as to his wealth and power—all these attributes and more can be applied either to the man or to the Society that is the subject of this history.

A skillful historian could, I suppose, describe a society's history chronologically, just as most biographies of individuals are treated, and produce a readable and cohesive story. I am impressed, however, by the fact that almost every change in the Society—its philosophy, operations, products, or activities—has resulted from slow evolutionary processes. Only two episodes in the entire 90-year history to date affected every single facet of the Society and its people at the same moments. These were the founding in 1888 and the Penrose bequest near the halfway mark in 1931. Even the Penrose bequest produced more changes in quantity than in kind; the Society expanded and expedited nearly all of its ongoing activities but introduced few new ones. Of these, perhaps initiation of the research-grants program and of the *Bibliography* were the most significant

new departures, but many others have followed through the years.

All other changes have cast their shadows before them for years or even decades, and few individual changes affected more than a small fraction of the total complex organism. For these reasons, I have chosen to treat the history topically. This calls for patience on the part of the readers, especially for one who might only be interested in what happened during a given year or decade, but at least the reader interested in certain topics can find them easily and as easily skip others.

The book has four parts. Part I tells of the founding of the Society, its growth to maturity, and how it came to inherit a fortune that has set it apart from most other scientific societies. Part II describes the evolution of the Society's makeup, its governance, and its administration, including the management and preservation of its fortune. Part III summarizes the activities, aspirations, and accomplishments of the Society—what it has done or tried to do through 90 years to advance the science of geology and how it has worked with its friends and neighbors toward that advancement. Part IV, the shortest and the most tenuous part by far, attempts to forecast the future of geology and of the Society that represents it. It also attempts to draw a few lessons from past experience. Several appendixes contain historical material that does not fit well into text format but that requires preservation for future reference.

The Geological Society of America in Profile

The Geological Society of America was founded in 1888 for the purpose of "the promotion of the science of

3

geology by the issuance of scholarly publications, the holding of meetings, the provision of assistance to research, and other appropriate means." These stated objectives have remained unchanged to the present day and will doubtless be continued far into the future.

Although retaining constant purposes, the Society has evolved and changed in innumerable ways since its inception—in size, in makeup, in ways of doing business, and in specific activities. Most of the changes were gradual, but one event that was to have far-reaching and long-continued effects occurred suddenly and almost without warning. This was the Society's inheritance in 1931 of a substantial endowment from a long-time devoted member and past officer, Richard Penrose.[1] By bequeathing half of his fortune to GSA, Penrose set the Society apart from virtually all other scientific and learned societies in North America. During its first 43 years the relatively small Society had depended on dues and subscriptions to finance all of its activities in meeting its objectives. With the Penrose endowment, it was suddenly able to expand not only in membership but also in all its activities and to undertake new ones as needs arose through the years. The Society's history is in reality a story of how it has managed a sizeable fortune in pursuing a single objective—the advancement of the science of geology.

As of January 1, 1981, the worldwide membership was 12,603; it consisted of 40 Honorary Fellows, 3,229 Fellows, 7,500 Members, and 1,834 Student Associates. Members are persons engaged in either geologic work (research or applied), the teaching of geology, or in graduate study of geology. Fellows are persons engaged in either geologic work or the teaching of geology who have contributed significantly to the advancement of the science. Honorary Fellows are persons distinguished for their attainments in geologic science in countries outside the North American continent. Prior to 1963 they were called Correspondents. Student Associates are full-time students majoring in geoscience at a degree-granting institution.

The management of the Society is the responsibility of the Council, which consists of four elected officers—President, Past President, Vice-President, and Treasurer—and twelve councilors, four of whom are elected each year to terms of three years. The Executive Director, who is appointed by the Council to do its bidding, serves as Secretary to the Council and the Society and is aided by a paid staff of forty to fifty people, some of them employed part time.

In 1980 there are within the Society six regional sections: Cordilleran, North-Central, Northeastern, Rocky Mountain, South-Central, and Southeastern. In addition, there are eight topical divisions: Archeological Geology, Coal Geology, Engineering Geology, Geophysics, History of Geology, Hydrogeology, Quaternary Geology and Geomorphology, and Structural Geology and Tectonics. Any member or Student Associate may affiliate with any or all of these sections and divisions.

The Society holds an annual meeting in the fall of each year with its associated societies. These include the Paleontological Society, the Mineralogical Society of America, the Society of Economic Geologists, the Society of Vertebrate Paleontology, the Geochemical Society, the National Association of Geology Teachers, the Geoscience Information Society, and the Cushman Foundation. Many other nonaffiliated societies and groups also meet with GSA on occasion. The divisions hold their annual meetings in the fall at the time of the Society's annual meeting. The sections of the Society, however, hold their annual meetings in late winter or spring.

Using the income from special endowments, the Society awards two medals: the Penrose Medal for "outstanding original contributions and achievements in the geological sciences" and the Arthur L. Day Medal for "outstanding distinction in contributing to geologic knowledge through the application of physics and chemistry to the solution of geological problems." Several of the divisions have also established awards to outstanding contributors to knowledge within their specialties.

The Society awards Penrose research grants for proposed investigations that will contribute to the purpose for which the Society was organized, that is, "the promotion of the science of geology." Such projects may be in support of doctor's or master's theses.

The Society sponsors the Penrose Conferences, the participants of which are all active researchers in diverse but related fields. These Conferences were initiated in 1969

---

1. Identity Note:—R.A.F. Penrose, Jr., and Spencer Penrose. Now that the Society's headquarters are in Colorado, more than ever before a special word as to the identity of GSA's principal benefactor is in order. Perhaps it will eliminate some of the questions asked by visitors to GSA in Boulder: Is this the Penrose who built the Broadmoor Hotel in Colorado Springs and the Manitou & Pikes Peak Railway, the one who made a fortune in Cripple Creek, who endowed the Penrose Memorial Hospital? and so on and on.

"Our" Penrose, Richard, is *not* the man whose name is almost a household word among Coloradans. That man was Spencer Penrose, a younger brother of GSA's R.A.F. Penrose, Jr.

Spencer was born November 2, 1865, and died December 7, 1939, eight years after his brother. He was more outgoing socially (some newspapers called him a "swashbuckler") and far more widely known than Richard, particularly in Colorado and Colorado Springs. He was also wealthier—in part because he was residuary legatee of several substantial family trusts that came down to him through his older brothers.

Spencer was basically a mining engineer with much of his practice in the Cripple Creek district of Colorado and was closely identified with ownership or management of several of the more fabulous mines of that superlatively productive district. He also joined his brother Richard in other profitable mining ventures in Arizona, at Bingham Canyon, Utah, and elsewhere, and was an officer or board member of several of the nation's great mining companies. He was an astute businessman and organized or managed innumerable businesses in the Colorado Springs area that fed on the product of the Cripple Creek mines. His philanthropies were great—both in life and after his death. From the myopic viewpoint of this history, however, the chief difference between Spencer and his brother Richard is that Spencer did *not* bequeath a fortune to the Geological Society of America.

and were patterned after the highly successful Gordon Conferences, well known arenas for the open and frank exchange of ideas in the field of chemistry.

The Society is by far the largest private publisher of geologic literature in the Americas. Its publications include the two monthly journals, the *Bulletin* and *Geology;* Memoirs; Special Papers; maps and charts; *The Treatise on Invertebrate Paleontology; Abstracts with Programs* for the section and annual meetings; the annual *Membership Directory;* a Society newsletter; and newsletters for all of the divisions. Until very recently GSA also published the monthly *Bibliography and Index of Geology.*

Official Seal and Unofficial Flag

Creation and use of a Society seal was authorized when GSA was first organized. According to Fairchild (1932, p. 204) the task of designing and preparing a Society seal was delegated by the Council on January 1, 1891, to the Publication Committee and the Secretary. The committee consisted of Grove Karl Gilbert, Henry S. Williams, and Charles H. Hitchcock; the Secretary was Herman Leroy Fairchild. The seal was completed and put to use by August 1891 on the program for the Fifth International Congress of Geologists, on Society stationery, and in Volume 2 of the *Bulletin.* Inasmuch as Gilbert, the committee chairman, and W J McGee, the Editor, were with the U.S. Geological Survey in Washington, it is reasonable to assume that the

design and preparation of the seal were accomplished by the Survey's skilled illustrators and engravers.

In gross aspects, the seal has remained remarkably unchanged through the years. It consists of an outline of North America, surrounded by a geologist's belt, with one variety of a geologist's hammer. The Society name and date appear on the belt. In detail, however, the seal has undergone many changes.

Minutiae lovers may be interested in studying random examples of the seal as reproduced from publications at various intervals (see accompanying figure). Note that the wrinkled belt smoothed out a few years after the Society became financially fat. Great Salt Lake has come and gone through the years, and various shadings have been applied first to the continent, then to the oceans, and at times to both. Type styles have evolved, and the ancient custom of separating words with periods has been abandoned. Depending partly on the background tone, the seal has been printed both in black on white and in reverse.

The names of the artists who designed the seal and made the many changes in it have not been recorded. In all probability various officers, staff workers, and printers tried their hands. To all of these unknowns and to Fred Handy, former GSA staffer and collector of the samples shown here, the Society owes its gratitude. In a way, the evolution of the seal is symbolic of the Society's own evolution—frequently changing in details to keep pace with the times but retaining the overall design unchanged.

1891–1935

1935–1945

1945–1959

1961–1972

1972–1979

For a short time in 1973 the Society actually had a semiofficial flag. Part of its story was told in a brief news item in *The Geologist* (v. VIII, no. 3), but the sequel was never publicized. A graduate student had received a small research grant to make underwater studies off the shores of Jamaica. Under joint sponsorship of the National Geographic Society, George Washington University, and GSA, the expedition involved a trip by the student and his wife in a small sailboat from the East Coast to Jamaica and return. He wished to fly the flags of all three sponsors during the voyage and asked GSA headquarters for its flag.

The Society had never had an official flag, and there are no records to suggest that one was ever considered. The staff accepted the challenge gleefully, however, and designed a flag (the Society seal in blue on a white ground), had it fabricated, and mailed it, on loan, in time for the trip. But the young couple met storms en route to Jamaica and their craft was wrecked, with loss of all gear including the flags. The crew and the remains of the boat were rescued by the U.S. Coast Guard. The young couple was divorced, the boat was sold in a division of property, the planned research was abandoned, and the Society was left without a banner. It has not been replaced and probably will not be.

Sources of Information

The most important single source of information for this compilation has been the set of official Council Minutes on file at GSA headquarters. These minutes vary widely in degree of detail provided, but all have the same disability: minutes are intended to record actions, not to report the facts or reasoning on which the actions are based. Nevertheless, the minutes of every meeting ever held by the GSA Council provide as complete a bare-bones record of all the Society's activities since 1888 as can be found anywhere. Flesh was put on the bones where needed by consulting the voluminous background material, such as committee reports used by the Councils in making decisions. These contain most of the facts, reasoning, and philosophies that were considered by both committees and Councils. Most of the more recent background material has been preserved and is on file at headquarters; it includes much correspondence, financial records, and other matter. The older backup records have suffered from attrition, and I used very few if any of them in this study.

Some of the facts recorded here came from my own memories or from those of various friends who have played large or small parts in Society history. These sources have been used sparingly, however, in part because memories are faulty. More importantly, even the oldest potential inform-

ants can have first-hand knowledge of only the latter half of GSA's story; these years tend to be better and far more completely recorded in sources other than the participants' minds.

More formally published sources of information than the minutes have been used liberally. Of these, Fairchild's (1932) history of the Society for the years 1888–1930 and the masterful biography of R.A.F. Penrose, Jr., by Fairbanks and Berkey (1952) were possibly the most fruitful, and much material herein has been lifted verbatim or paraphrased from them.

Other published sources, used mainly as backups to the Council Minutes, include the *Proceedings, GSA Annual Reports,* which contain reports of the officers, *Membership Directory,* and the like; all of these are permanently available in many geological libraries. The Society's correspondence, accounting, and similar files have been consulted sparingly, but they constitute a rich source of detail for future historians. Headquarters has not maintained a central file for many years; the records must be sought in the files of the various departments. Preserved records vary widely, both quantitatively and qualitatively, depending on the judgments of successive department heads and their secretaries.

Doubts as to the propriety and desirability of leaning so heavily on Council Minutes for this history have beset me throughout my work on it. Fairchild (1932, p. 212) says that he avoided the minutes as primary sources for his history (though he must have had almost total recall over a long lifetime to have recovered many of the facts he did without consulting the minutes). Fairchild's reluctance to use the minutes was based largely on his belief that these and their backup material were to be considered as "confidential" and for the use only of the Council. Most such records are indeed so marked at the time of distribution and first use. Moreover, some of the records contain material that might damage reputations or egos and embarrass the Society or individuals if they were made public. I have tried to be discreet about use of such material and have either discarded it or resorted to generalities and exclusion of identities. Other than this, I have chosen to draw freely on the minutes and to quote or paraphrase wherever necessary. This decision was based on my firm belief that the Society and all its records belong to the entire membership; hence the records cannot be legally or morally withheld from GSA members or the public. True, the records are seldom consulted by the membership or others, but they are openly displayed in the conference room at headquarters and may be studied by any Society member who visits that building.

2

Birth of the Society[2]

Status of Geology in 1888

When the Society was founded in 1888, geology was fairly well established both as a science and as a respectable way of life for its practitioners.

The principles and many of the details of stratigraphy and paleontology were well known, particularly in the northeastern part of the United States. The record of ancient life as revealed by fossils had been fairly well delineated, but the origin and extinction of species were still subjects of speculation. Quiet deposition of sedimentary rocks in bodies of water or even on land surfaces was favored by almost everyone over their formation by gigantic floods. The principles of physiography (now geomorphology) and with them the recognition of the role of glacial ice as a sculpturing agent were well established. Many elements of structural geology and of mountain building were still enigmatic, but good beginnings had been made in deciphering the sequence of events that had transpired during the formation of certain mountain systems. Aided by the polarizing microscope and its application to thin sections, igneous rocks and their component minerals were beginning to be well understood. Topographic maps and cartographic methods had been developed to a degree that had permitted production of useful geologic maps in greater or less detail of large parts of North America.

In large part, geology was still an intellectual exercise

as it had been for more than 2,000 years, pursued to satisfy scientific curiosity rather than to serve practical ends. The great surveys of the western territories had produced masses of new scientific and practical knowledge, but the chief purposes of most of them were exploration and physical inventories of unknown lands. Sound beginnings had been made in what was to become known as economic geology. Several superb studies of individual ore deposits or of entire mining districts had already laid down a sound methodology for such studies to which we still adhere. Application of geology to study of ore deposits or for any other practical purposes, however, was in its infancy. Many uses of geologic knowledge that we take for granted today were unthought of, as were innumerable facts and theories about the character and makeup of the earth itself. Some of these awaited discovery only for lack of time, patience, and finances to amass more observational details; others awaited invention of new tools—geophysical instruments, aerial photographs, deep drilling apparatus, radioactivity, X-ray apparatus, and isotopic and other age-dating techniques, among many.

Exact numbers are not known, but in 1888 there were not many more than 200 geologists at work in all of North America. Nearly all of these were in the colleges and universities, a number of which had already established departments of geology, or in the state or federal geological surveys. Those in educational institutions did independent research in addition to teaching the science, or else worked part time for one or another of the governmental surveys. Then, as now, satisfaction of intellectual curiosity was more important to most than were financial rewards; those with

---

2. This chapter is based almost entirely on Chapters I through V of Fairchild's history, with much condensation of his material and perhaps some differences in emphasis that result from another half century of perspective and a different author. For more details and for a thumbnail sketch of the development of the science from 614 B.C. onward, the reader should turn to the original document (Fairchild, 1932).

private incomes had far more freedom than others to pursue scientific truths. Except as knowledge of geology and mineralogy was one of many skills that mining engineers must possess, there was no such thing as staff or consulting work for companies, such as now supports thousands of professional geologists.

Though their numbers were few, some of the best minds the geologic community has ever known were in the ascendant. A few such had gone before and had laid the foundations, and many were to come later, but the geologists of the latter half of the nineteenth century were building, stone by stone, a lasting temple of science on those early foundations. These men were explorers not only of the lands they surveyed but also of the universe. Only a few had highly specialized interests; most were natural scientists in the broadest sense, to whom emphasis on geology represented in itself a high degree of specialization.

Early contributors to the literature of geology published their papers privately in the *American Mineralogical Journal* (1810–1814), the *American Journal of Science* (1818 onward), or in the transactions of the broader philosophical societies or academies. (Thomas Jefferson, America's first vertebrate paleontologist but better known for certain other accomplishments, published his first and only paper on fossils in 1797 in the *Transactions of the American Philosophical Society* [Jefferson, 1799]). Beginning about 1830, state geological surveys soon established their own reports series, some of them illustrious, and furnished additional outlets. The third oldest national geological survey in the world, the Geological Survey of Canada, was established in 1847; its publications are continuous from that date. Within the United States, the Hayden, King, Powell, and other exploratory surveys of the West each had its own governmental publication outlet. Their successor, the U.S. Geological Survey (USGS), established in 1879, was already well on its way to pre-eminence as a publisher of geologic maps and reports by the time GSA was formed nine years later. Like the state surveys, however, the national surveys could publish only the work of their own staffs. This left academic and independent geologists with no North American journal devoted specifically to geology. Fulfillment of their publication needs was one of the more compelling reasons for formation of GSA.

Ancestors, Indirect and Direct

GSA (1888) is directly descended from the Association of American Geologists (1840) and the American Association for the Advancement of Science (AAAS, 1848). Several efforts had been made earlier in the nineteenth century to establish national societies of geologists, but none of these left direct connections to the Association of American Geologists. The first of these, the American Geological Society, organized by William Maclure in 1819, lived for only

ten years, with no descendants. The Geological Society of Pennsylvania, organized in 1832, had broader aspirations than the geology of a single state, but it survived for only a year or two.

The Association of American Geologists was the first quasi-permanent national geologic society in the United States. It was organized formally at Philadelphia on April 2, 1840, but had its beginning at the home of Ebenezer Emmons in Albany, New York, in 1838–1839 as the New York (State Geological Survey) Board of Geologists (a small group of what would be called Regional or District Geologists in a state organization today). This group soon saw the need to draw in the members of other state geological surveys, particularly those of Pennsylvania and Massachusetts, with the primary aim of unifying Paleozoic stratigraphic nomenclature. The Association of American Geologists, soon broadened to the Association of American Geologists and Naturalists, was the result. It failed in meeting its primary objective but did hold meetings annually through 1847 and published its proceedings in the *American Journal of Science*.[3]

In 1848 the Association broadened its interests still further and became the American Association for the Advancement of Science (AAAS). Thus the foundations of the AAAS, which stands today as the world's largest and most all-inclusive group of scientists, were laid by American geologists. Much later, in 1888, the tables were turned, and the geologists' offspring became the parent of the Geological Society of America.

The American Association of Geologists and Naturalists lost its identity when the AAAS was formed, but geologists retained strong voices in the new society, ten of them becoming presidents in the ensuing 40 years, and others becoming organizers and leaders in what finally became known as Section E (Geology and Geography) of the AAAS.

Founding the Geological Society of America

The seeds for what was to become GSA were sown at Cincinnati, Ohio, in 1881. There, during the annual meeting of AAAS, a small rump group of geologists met to consider the advisability of establishing a separate geological society. The Winchell brothers, Alexander and Newton, with Charles H. Hitchcock, seem to have taken the lead; no complete list of other attendants was kept. Newton Winchell chaired the first meeting, but Alexander assumed the lead in subsequent ones. Similar rump meetings were held at AAAS meetings in Montreal in 1882 and at Minneapolis in 1883 with some changes in participants.

During this same period, many of the same geologists

---

3. Reports for the 1840, 1841, and 1842 meetings of the association were also published in book form by Gould, Kendall and Lincoln (Boston) in 1843. This work was reprinted in 1978 by Arno Press (New York).

Newton H. Winchel, 1839–1914. President, 1902.

Alexander Winchel, 1824–1891. President, 1891. Leading organizer and founding father of the Geological Society of America, 1881–1888.

interested in a new society were also playing parts in the formation of an International Congress of Geologists, later the International Geological Congress. The political infighting that resulted, which also involved the AAAS, the U.S. Geological Survey, and other groups in North America and elsewhere, no doubt diluted and delayed some of the efforts toward an American society. In any event, the proposal to form a separate organization of geologists lay dormant, at least on the surface, until 1888.

These early efforts to establish a separate society for geologists were motivated by several factors. Among the stated reasons were:

1. The rapid growth of geology as a science.

2. Diffusion of geologic knowledge through many literature sources.

3. Dissatisfaction with AAAS as a medium for exchange of geologic thought because (a) meetings were held in summer, the geologists' field season; (b) AAAS meetings were more social than scientific; (c) lack of specialized pub-

lication facilities; and (d) AAAS interests were so broad as to preclude space for specialties.

4. Geologists, as a body, had no way of expressing opinions on geo-political issues except through AAAS, whose meetings were not well attended by geologists.

5. America needed a geological society and a geologic journal, such as those that characterized the geologic sciences in Europe (abstracted from Fairchild, 1932, p. 65–66, in his quotation from Winchell, 1914).

The proposed break from AAAS, together with the expressed and implied criticism of that organization, raised doubts in many geologists' minds, particularly among some of the older conservatives. These doubts and delays in their resolution despite what must have been a determined lobbying campaign by the leaders accounts for the slow progress and long delay in positive and overt action.

By 1888, the organizers must have felt that doubts and oppositions had been dispelled and that the time for action had arrived. A new journal, the *American Geologist,* which had been conceived during the rump meeting at Montreal in 1882, made its first appearance in January 1888.[4] Its June 1888 issue contained a "Call" to all American geologists to meet at Cleveland, Ohio, on August 14, 1888, the day preceding the AAAS annual meeting (Winchell and Hitchcock, 1888). The purpose of the meeting—to organize an independent geological society—was stated forthrightly, as were these obviously politically motivated limitations:

1. Members of the new society must also be members of AAAS.

2. Officers were to be the same as those elected by AAAS as President and Secretary of Section E.

3. AAAS was to be asked to allow Section E to hold its meetings separately and independently from those of the parent organization.

---

4. This journal published 36 volumes in the next 18 years, paralleling GSA's *Bulletin*. In 1905 it merged with the newly established journal *Economic Geology* and went out of business as a separate entity.

In short, the new society was to be little different from Section E of the AAAS, but with a new name.

The Cleveland meeting was held as scheduled, in harmony and with enthusiasm. Because they really represent the birth certificate of the Geological Society of America, the minutes are reproduced here.

The progress made at the Cleveland meeting and the gracious acquiescence of AAAS itself led to renewed, almost feverish, activity on the part of the organizing committee, guided largely by Alexander Winchell and John J. Stevenson, who was to become the new Society's first secretary. Within three months after the Cleveland meeting, three circulars to geologists were issued outlining plans for an American Geological Society, presenting a provisional constitution and set of bylaws, and seeking firm commitments from prospective members. A ballot for nomination of officers and for election of Fellows also went out with the final circular, and the formal organizational meeting of the new society was called for December 27, 1888, at Cornell University, Ithaca, New York. Already, be it noted, was the Society breaking away from AAAS by calling a completely independent organizational meeting.

The membership requirements established by the organizing committee seem unduly elaborate today, but they were evidently dictated by the sensitivities engendered by the proposed break from AAAS. In essence, there were to be two classes of Fellows at the start. The first of these, to be called Original Fellows, was to consist of all working geologists and teachers of geology *who were also members of AAAS* and who declared an interest in joining and in paying $10 per year in dues. The second class was to include any other geologists who were willing to accept the conditions of fellowship, *were not members of AAAS,* and were elected by the members of the first class. As soon as the total of the two classes of applicants reached 100, the Original Fellows (class 1) were to vote on applicants of class 2, then all the 100 geologists were to become founders and to elect officers and adopt a constitution. Incidentally, the secretary of the organizing committee admitted much later that the annual dues were set at $10—an exorbitant sum in 1888—in order to ward off applications from nongeologists to become original members (Stevenson, 1914, p. 15). The requisite number of applicants (actually 98, rather than 100) was reached by October 1888, and the last circular contained the list, together with a formal call for the organizational meeting at Cornell.

The first GSA *Bulletin* (v. 1, p. 579) lists 12 Original Fellows, plus 14 others (all non-AAAS members, and including 2 Canadians) who were elected on December 27, 1888. This same list shows that by the end of 1889 a total of 191 Fellows had been elected, although 4 had already died.

Fairchild's discussion of the early membership (1932, p. 97–106) and his attempt to do honor to the Original Fellows as distinct from those who were admitted later, is heartwarming and praiseworthy but also confusing. Most of the original records still exist at headquarters, are mentioned in Fairchild's book, or have appeared in the *Bulletin* from time to time, but some early applicants withdrew or were not elected, some failed to pay their dues, and attrition took its toll of the lists. Fairchild himself points out that the list of 112 Original Fellows contains several errors of commission and omission. The best I can do is to say that about 100 geologists had qualified for membership by the close of the Ithaca meeting and were thenceforth carried on the rolls of the Society as Original Fellows. The last of these, Nelson Horatio Darton, died in 1948. Among the founders were the majority of the leaders in the science in the United States—the outstanding teachers, the early explorers of the West, the directors and staffs of the state and federal geological surveys, and many others.

For some reason now unknown, no Canadian geologists were among the Original Fellows. They already had their own Royal Society, but many of them belonged to Section E of the AAAS, and they must have received the same invitations as did their United States colleagues. Whatever the reasons for hesitation may have been, they dissipated quickly for two Canadians were elected by the Original Fellows group in 1888, and by the 1889 Annual Meeting in Toronto, a dozen of the outstanding geologists of Canada had been elected to Fellowship. Ever since, Canada has been strongly represented in the membership and management of the Society. The Society was well named from the beginning—it is the Geological Society of America, not the Geological Society of the United States of America.

The organizational meeting at Ithaca went smoothly and accomplished its objectives. The minutes appear in full in Volume 1 of the *Bulletin,* and in Fairchild (1932, p. 88–92) and are not reprinted here. Thirteen Fellows were present—a far higher percentage of the total membership than appears today at the Society's annual business meetings! Fellows were elected, plans were made to revise the tentative Constitution which had just been adopted, a Publications Committee was appointed, and the usual minor business items disposed of. Officers were elected for 1889 as follows:

President—James Hall, Albany, New York

First Vice-President—James D. Dana, New Haven, Connecticut

Second Vice-President—Alexander Winchell, Ann Arbor, Michigan

Secretary—John J. Stevenson, New York, New York

Treasurer—Henry S. Williams, Ithaca, New York

Members-at-Large of the Council—John S. Newberry, New York, New York; John W. Powell, Washington, D.C.; Charles H. Hitchcock, Hanover, New Hampshire

Seemingly, Alexander Winchell should perhaps have been the first President, if only because he had been the most active and influential person in the movement to establish a society. Moreover, he was widely known as geologist and administrator, having served as President of Syracuse University and as Professor of Geology at the University of Michigan. But as Fairchild says, "Dr. [James] Hall was a dominating personality, the nestor among American geologists; the meeting was in his State, and on his 'stamping ground' in the early New York survey, and so he was made the first President and the only President of the 'American Geological Society'" (Fairchild, 1932, p. 153).

Winchell swallowed gracefully whatever natural disappointment he may have had and accepted election as Second Vice-President. He was elected to the presidency for 1891, but unfortunately did not live long enough to preside in his own right over his brainchild. He did, however, have a chance to lead the Society for a full year in 1890 during the continuing illnesses of James D. Dana and

James Hall, 1811–1898. First President of the Geological Society
of America, 1889.

John S. Newberry, the President and First Vice-President
elected for that year. A similar situation developed in 1950
when Chester Stock, President-elect for 1951, died shortly
after his election. Thomas S. Lovering, Vice-President,
acted in Stock's place during 1951 and then became President
for 1952.

The name "Geological Society of America" was not
formally used until the end of 1889 when the revised Constitution
was adopted. Before that, the informal name was
Association of American Geologists, which soon became,
more formally, the American Geological Society. Perhaps
these names were used only in further deference to the parent
organization because of similarity to that of the American
Association for the Advancement of Science. The
change in nomenclature to Geological Society of America
was accepted informally at the organizational meeting in
Ithaca, but formal adoption was delayed, apparently for
parliamentary reasons—the Ithaca meeting was only accepting
a provisional Constitution that used the earlier
name. But the group at Ithaca clearly indicated its intentions
when it authorized Fellows to follow the English and

European custom by use of the initials F.G.S.A. after their
names.

Professor Alexander Winchell summarized the accomplishments
of the Ithaca meeting and the vision of the
founders thus (Fairchild, 1932, p. 93):

The Geological Society thus began its existence strong in
numbers, ability and finances. It had already enlisted the adhesion
of almost every working geologist in the United States, and none
unworthy had been permitted to enter. Thus was established again
an authoritative representative of American geology, competent
to know what the interests of American geology demand, and with
full liberty to act from motives lying exclusively within its own
field. May peace and a spirit of mutual consideration, sympathy
and helpfulness reign within its borders. May the wise counsels of
its first President remain a testament to guide the footsteps of
many generations in the ways of usefulness and honor.

Thus was GSA born. It held a summer meeting in
Toronto in August 1889 at the close of the AAAS meeting
and worked over a proposed revision of the Constitution; a
few scientific papers were read. GSA's second annual meeting,
and the first to emphasize presentation of technical
papers, took place in New York City on December 27,
1889. The greatest accomplishment at New York was formal
adoption of a revised Constitution and Bylaws that
were to survive almost intact for many years (Appendix A).
By the end of the Society's first year, in December 1889, the
membership had nearly doubled, to a total of 191 elected
Fellows, though 4 of these had died, leaving an active
membership of 187 (*Bulletin*, v. 1, p. 579).

At the 1888 organizational meeting in Ithaca, officers
were elected in open meeting from a list of nominees who
had been suggested by potential members. For the next
several years, elections took place in open meeting, or by
written ballot from Council-prepared slates of three names
for each office. The rules were changed in time for the
election of 1895 officers, and single slates of nominees have
been in use ever since. Dissatisfaction with the single-slate
system has risen from the electorate from time to time, and
two such slates, those for 1897 and for 1922, were actively
contested.

Charles H. Hitchcock, who had been second only to
the Winchells in his active devotion to the cause of an
independent Society, was the only presidential candidate
ever to be defeated in an election. Fairchild tells the sad
story with sympathy and feeling (p. 156). Hitchcock was a
member of the Council in 1889 and 1890 and later was
made Second Vice-President for 1895 and First Vice-President
for 1896. When he was nominated for the presidency
for 1897 according to custom, however, an opposition
slate was nominated. It was sponsored by nine of the older
and more eminent Fellows, three of them Past Presidents.
The opposition slate won the election; Hitchcock was de-

feated and somewhat discredited among his peers. In retrospect, the only reason that Fairchild could decipher for the opposition was that Hitchcock's scientific accomplishments over 40 years ". . . offered no features or phenomena of striking interest with discoveries of popular appeal."

Again, in 1921, an independent ticket was sponsored by 68 Fellows. This one, however, was heavily defeated at the polls. Defeat was almost guaranteed in advance because the candidates nominated by the opposition group were unwilling to muddy the waters and to hurt feelings by allowing use of their names. Through parliamentary maneuvers, their identities finally came before the electorate, but the overwhelming majority vote was in favor of the original ticket.

From Infancy to Maturity

From its birth in 1888, with a rich inheritance of its parents' dreams and aspirations, the Society grew slowly toward maturity. By 1930, some 42 years after its founding, it had affirmed its place of leadership in the development of geologic thought in North America. This it had done primarily through publication of scientific papers in a quarterly *Bulletin* and by the holding of annual meetings for oral presentation and free discussion of research results in virtually every facet of geology. Always self-sufficient financially, it had accumulated an impressive nest egg of investments; its annual expenses, principally for publication, were balanced by income. Leadership and management of its affairs were done entirely by volunteers drawn from the membership, which had grown from the original 100 or so to about 600.

The Society had already produced strong and healthy offspring—one subdivision and several new, separate, and more specialized scientific societies. The still vigorous parent was to produce more in the years to come. Happily, all of the offspring are still on compatible and friendly terms with the progenitor and join with it regularly for family reunions.

During these 42 years GSA had seen enormous growth in the geologic population and with it development of other, unrelated societies, each with its own publication series. With all of these, as well as with the growing governmental geological surveys and their publishing houses, the Society continued to exist on terms of helpful cooperation rather than of competition. Eligibility to GSA fellowship, originally almost a right derived from one's vocation, had early become an honor to be coveted by those aspiring to leadership in the science. Multiple memberships, therefore, with allegiance to GSA as representing the broad spectrum of geology and to one or more specialistic societies had become a way of life for many geologists.

All of the facets of GSA's growth from infancy to maturity are recounted in the following pages. It is necessary to pause here, however, for discussion of a single event that was to set GSA aside from all other geological societies. Described in the next two chapters, this event was the Society's inheritance of nearly $4,000,000 from one of its loyal members, R.A.F. Penrose, Jr. By his generous gift effective on his death July 31, 1931, Penrose made it possible for GSA, already mature and with well-defined ideas and ideals as to how it might advance the science of geology, to do far more toward that advancement in the years to come than it could ever have done on its own.

3

Richard Alexander Fullerton Penrose, Jr., the Man and the Benefactor

The Man and His Background

What kind of man was R.A.F. Penrose, Jr., and how did he come to choose this Society—and the American Philosophical Society—to be his heirs? Here are the dry facts that Penrose himself thought worthy of permanent record. They are taken verbatim from the same page of *Who Was Who in America,* Volume 1, that carries biographies of his father, three of his brothers, and several other relatives.

Unaccountably, Penrose did not list his affiliation with the American Philosophical Society nor with GSA. Nor did he mention that he was GSA President in 1930, an honor which he was reported by contemporaries as having considered the greatest of his career. These omissions may merely reflect delays in updating of his autobiography.

Throughout this book I use the name Penrose or Richard where it is necessary to distinguish between him and his father, R.A.F. Penrose, or his brothers. This is out of no disrespect to him. Penrose himself seldom used his full name, Richard Alexander Fullerton Penrose, Jr. He was too modest to use the term "Doctor" and seems to have preferred that others not use it in addressing him. He signed his business letters as "R.A.F. Penrose, Jr." or simply as "Penrose." Family letters he commonly signed as he did his business letters or, more rarely, used his initials, "R.A.F.P., Jr." or "Dick." Unlike many men who drop the term "Junior" after their fathers' deaths, he retained it throughout his life. This preference was partly out of respect for his

PENROSE, Richard Alexander Fullerton, Jr., geologist; b. Phila., Pa., Dec. 17, 1863; s. Richard Alexander Fullerton and Sarah Hannah (Boies) P.; A.B., Harvard, 1884, A.M., Ph.D., 1886; unmarried. Geologist in charge survey of Eastern Tex., for Tex. Geol. Survey, 1888; apptd., 1889, by Geol. Survey of Ark. to make detailed reports on the manganese and iron ore regions of Ark.; asso. prof. economic geology, 1892-95, prof., 1895-1911, U. of Chicago. Lecturer on economic geology at Stanford, 1893; spl. geologist U.S. Geol. Survey, 1894, to examine and report on gold dists. of Cripple Creek, Colo. Mem. bd. mgrs. Phila., Germantown & Norristown Ry. Co.; dir. various mining corps., also Ridge Avenue Passenger Ry. Co. of Phila. Trustee U. of Pa., 1911-27; pres. Acad. Natural Sciences of Phila., 1922-26. Asso. editor Journal of Geology. Mem. Soc. Economic Geologists (pres. 1920-21); mem. Nat. Research Council (geology com., 1917-18, div. of geology and geography, 1919-23); del. Internat. Geol. Congress, Stockholm, 1910, Toronto, 1913, Brussels, 1922, Madrid, 1927; mem. Fairmount Park Commn. (Phila.). Republican. Author: The Nature and Origin of Deposits of Phosphate of Lime; Geology of the Gulf Tertiary of Texas; Manganese: Its Uses, Ores, Deposits; The Iron Deposits of Arkansas; What a Geologist Can Do in War; The Last Stand of the Old Siberia. Home: Philadelphia, Pa. Died July 31, 1931.

father and partly to avoid confusion in estate settlements and other legal papers.

More facts than given in *Who's Who* are needed to

14

flesh out the bare bones of this biographical sketch of a remarkable man. The remainder of this chapter is based on the exhaustive and fascinating Penrose biography (Fairbanks and Berkey, 1952), plus gleanings from Penrose's own memoirs of his early years, from several of many published memorials, and from his own notebooks, diaries, and letters preserved in GSA archives. For details of his life, and perhaps even for an entirely different picture of the man, the reader should consult these same sources. This is but a brief summary, colored by my own second-hand impressions of a highly complex person and his career.

Penrose was born into a patrician family whose forebears came to America from England about 1700 to join William Penn's colony in Philadelphia. The family had been more than comfortably situated financially and outstanding in political, military, industrial, and scientific fields, not only for six generations in America but also for many generations before that in England, first in Cornwall and later in Avon.

Richard was the fourth of seven sons. He and three of his brothers, Boies, Charles, and Spencer, lived to achieve maturity, fame, and wealth in their own rights. Only Charles and Spencer ever married, and only Charles had children.

The Penrose family always lived comfortably, though modestly and without ostentation. Much of the family wealth was in trust funds, the income of which went to each of the sons as they grew up and then to their surviving brothers as each died. Though none of them could touch the principal, this arrangement meant a lifetime of comparative financial freedom for Richard and his brothers. More important, it meant ready sources of money for his early investments. Details are lacking, but from passing mention in many family letters it appears certain that his father, as well as brothers Boies, Charles, and Spencer, depended on him for investment advice, particularly as to mining ventures.

In his memoirs, Penrose says, "I will not dwell on my childhood, for it was doubtless the same as that of millions of other boys." In this, Penrose was characteristically overmodest. Comparatively few children of the post–Civil War epoch had their primary and secondary education in private schools or with tutors in their homes. Fewer still went to Harvard at 17, obtained an A.B. *summa cum laude* at 21, and the Ph.D. at 23. And not many college students have lived with their brothers in a house bought for them by their parents, complete with an aging maiden cousin as housekeeper, rather than sharing dormitory and dining hall life with other students.

Penrose was an earnest, hard-working, and highly successful student, as shown by his *summa cum laude* and his election to Phi Beta Kappa. More than most students, he knew exactly why he was in college and how to get the most from it. A tall, imposing man, he cared for his physical and mental development, and while a graduate student found time and energy to stroke the Harvard crew to victory over Yale. Through most of his life he was an active and successful big-game hunter and fisherman.

The boys' mother, Sarah Hannah Boies Penrose, died in 1881 when they were in their teens. In consequence, the senior Penrose devoted much of his remaining life to being both father and mother to his children. In this he was obviously successful, launching each on a career of his own. Some might think he was oversolicitous, if not domineering, but all his sons seem to have responded well to his guidance. Richard, in particular, who was clearly the favorite son, was devoted to his father and dutifully obeyed his every wish until his father's death in 1908, when he himself was middle-aged. While in college, Richard sent drafts of all his term papers home for parental review and correction before submitting them, and he maintained a warm and constant correspondence with his father during travels. He acknowledged fatherly advice courteously and apparently followed it, whether it had to do with health maintenance, bathing habits, or simply demands for weekly letters. In addition, of course, he kept his father fully informed as to his activities or his impressions of countries or fellow travelers.

He was intensely loyal to his father's wishes and to inherited family sentiment. For example, he maintained the family home in Philadelphia, complete with staff, until his own death, although he himself lived in a hotel suite and the home was seldom used by him or any other family member after his father's death. (Although he did not live there, most of his family and business correspondence was handled at an office in the home by his long-time confidential secretary rather than in his main office.)

Again, and because "he is the only one of my sons who has taken any interest in my place" his father bequeathed him the beloved summer home "Sea Bench," at Seal Harbor, Mt. Desert Island, Maine. The will expressed a wish that the place be kept in the family for at least a time after Dr. Penrose's death, and Richard did not sell it until 1927, although it was rarely used by him or other family members.

His father's death in 1908 affected him deeply. Coincidental or not, the year also marked a radical midpoint change in Richard's professional career. Whether this change in career objectives and life style had to do with grief over his loss of a father for whom he had deep affection, with loss of incentive and domination from a source outside himself, or with other causes is something perhaps a psychologist could answer.

Professional Career, First Half

From 1884, when he left Harvard, until 1908 Penrose was an ambitious and serious scientist, eager to advance his knowledge and stature in his chosen profession. His interest seems to have been always on the economic potentials of his study area, but he willingly undertook mapping of Tertiary sediments of the Texas Gulf Coast, under primitive conditions, and produced excellent reports on them. He moved from the Texas Survey to that of Arkansas, not so much for a change of geologic scene, but because he believed he could further his own geologic career by working under J. C. Branner, then State Geologist.

Penrose's early field notebooks, many of which are preserved at GSA headquarters, reflect a careful observer and recorder of facts, blessed with extraordinary judgment. All notes are clearly written, mostly in whole sentences. Some, such as measurements of sections or traverses, were evidently made on the ground; others seem to have been done at the end of the day's work (but these were obviously summarized from some kind of notes, not now preserved, that were made at the time of observation).

The most striking fact about his field notes, striking because the habit is so rare among contemporary geologists, is that they contain excellent first drafts of what were to become final reports or recommendations to his private or public employers. One notebook on some mine examinations in Idaho, for example, even contains a draft prospectus for public sale of stock to finance a large exploration and mine management company.

These partial drafts characteristically appear in ink on the backs of pages that contain the original penciled field notes. The last couple of pages of many notebooks contain meticulous lists of field and travel expenses for the work reported on earlier pages. Interestingly, items for "whiskey," "brandy," or more cryptically, for "expense," commonly exceed the charges for meals and rooms.

Through 1907, at least, Penrose also kept diaries of his travels, particularly those abroad. Taken together, they constitute a brief but clear and objective description of the physical geography of many parts of the world; only rarely did he digress to observations of human behavior or of social geography.

He visited most of the world's great mining districts and personally examined many individual mines. These examinations were cursory, but he came from each with good understanding of the kinds of things that a mining geologist—or a potential investor—wants to know, with emphasis on size and tenor of ore bodies, mining methods and metallurgical problems, and but little speculation as to origin, detailed mineralogy, and the like. Wherever he went, Penrose seems to have gone armed with letters of introduction to the most knowledgeable and influential mine owners, managers, or others in authority. These letters led to personally guided tours and frank conversations; and, together with a keen observational ability and a thorough knowledge of the literature, they must explain how he was able to amass so much significant information in

such brief visits to entire countries or to their mineral deposits.

He did many things during those first 24 years, studying the Cripple Creek ore deposits for the U.S. Geological Survey (Cross and Penrose, 1895), examining innumerable mines for clients or on his own behalf, lecturing on economic geology at Chicago and at Stanford, editing the *Journal of Geology,* and traveling throughout the world. He also began the investment career that was to dominate his later life. He refused to join his brother Spencer in Cripple Creek district investments because he felt bound by the honor code of the U.S. Geological Survey, which had paid him to study the mines. But during the nineties he made his first, and highly successful, investments in several great silver and copper mines in Arizona. In 1903, Richard and Spencer joined with Daniel C. Jackling and three or four others to found the Utah Copper Company formed to develop the fabulous Bingham Canyon porphyry copper mine in Utah. Utah Copper, now part of Kennecott, prospered from the start and was the basis of Penrose's entire fortune. He sold his interest in 1925.

Professional Career, Last Half

The last half of his career, beginning in 1908, was different from the first half. Always shy and withdrawn, he became even more so, especially during the last few years of his life when he became more and more reclusive and almost antisocial.

The careful recording of geologic and other observations and of travels in notebooks and diaries ceased abruptly in 1908, and the files he left at death contain curiously little correspondence or other records of his activities, investment decisions, or other matters. He published almost nothing in the scientific literature. This is not to say that Penrose had "burned out" professionally at 45. Quite the reverse was true, but he had, consciously or not, ended the scientific learning stage of his life and had entered one of consolidation and of cashing in on his knowledge. He continued to travel extensively but more on business than for relaxation. He continued to examine ore deposits but more as potential investments for himself or others than as geologic curiosities. More and more of his life seems to have been devoted to making and managing investments.

Few records exist of these activities, but the size of his fortune is ample proof that he was remarkably wise and successful. As officer or director of innumerable companies, not only in mining but in many other fields, he must have taken leading parts in their guidance, for he seems never to have accepted a job, whether for a company or for one of his many societies and clubs, without devoting the best of his energy and wisdom to its betterment. He had few close personal friends and tended to avoid them when he could, but he had numbers of close business friendships among the great and near great. Partly as a result of this and partly because of family connections, he was showered with honors. He was elected to dozens of societies and clubs more often than not attaining high office in them or at least being offered election to office. He remained active to the end, in investments, as board member of numerous corporations, and in his extracurricular work for his societies.

Always generous, he became even more so as he grew older, bestowing many thousands of dollars on a wide variety of recipients. Though generous almost to a fault to any cause in which he believed, he was also discerning and never hesitated to refuse help to any request that he felt to be unworthy or on which he was importuned too strongly. In turning down a payment, he was always courteous, commonly pleading pressure of other needs for his help, but he was firm in his denials.

During his final years, as if in anticipation of death, he became even more reclusive and devoted much of his attention to a search for an heir. This search is described below. After only a brief illness, death came quietly; it was ascribed to chronic nephritis and arteriosclerosis (*New York Times,* August 1, 1931).

The Penrose Bequest

Penrose died July 31, 1931. Exactly a week later his will was offered for probate in the Orphans Court of Philadelphia. So far as can now be determined, its publication brought the first real confirmation to anyone, heirs or others, as to the disposition of his estate. Not only was the estate much larger than would have been expected by most, but the bulk of it went to two learned societies—the American Philosophical Society and the Geological Society of America.

As early as 1928, GSA Secretary Charles P. Berkey, if no one else, surely knew something of Penrose's plans to support the Society. In his 1927 report, Berkey told the Council that he wished to relinquish the Secretary's job at the end of 1928. He also warned that the Society would soon need a paid Secretary. Yet the following spring at the Council meeting of April 14, 1928, he abruptly withdrew his plan to resign, saying, "A new factor has been injected into the problem, which contemplates a considerably expanded support for the operations of the Society and perhaps a good deal of reorganization. This plan has not reached a point where it can be presented yet, but it seems promising enough to warrant delay on the secretarial matter" (Council Minutes for 1928, p 310). Penrose had not yet written his final will in 1928, but in view of later events, it seems almost certain that he had some plan to endow the Society and that he had told Berkey of this plan.

The will, written in the testator's own hand, and dated June 12, 1930, is quoted here in full, minus only the signature and the initials on each paragraph that made it a legal instrument, (Council Minutes for 1931, p. 85; Fairbanks and Berkey, 1952, p. 752–753).

Be it remembered that I, Richard A. F. Penrose, Jr., of the City of Philadelphia and the State of Pennsylvania, do make and declare this as and for my last will and testament in the manner and form following, to wit:

First: I give, devise and bequeath to my brother, Spencer Penrose, all my right, title and interest in the property over which I have the power of appointment conferred on me by the will of my father, Richard A. F. Penrose. I also give, devise and bequeath to my said brother, Spencer Penrose, all my right, title, interest and estate whatsoever arising under the will of my mother, Sarah H. B. Penrose, and I also give and devise to my said brother, Spencer Penrose, all my right, title, interest and estate in the property known as 1331 Spruce Street in the City of Philadelphia and the State of Pennsylvania.

Second: I give and bequeath to each of my three first cousins in Carlisle, Pennsylvania, namely Miss Sarah M. Penrose, Miss Ellen W. Penrose and Miss Virginia A. M. Penrose twenty thousand dollars ($20,000).

Third: I give and bequeath to my Secretary, Miss Marion L. Ivens, twenty-five thousand dollars ($25,000) provided she is in my employ at the time of my death. I give and bequeath to my assistant Secretary, Miss A. M. Feeney, three thousand dollars ($3,000) provided she is in my employ at the time of my death, and I also give and bequeath to the assistant stenographer or clerk who may be helping her with my work at the time of my death five hundred dollars ($500).

Fourth: I give and bequeath to Thomas Tobin who was once in charge of our old stable on Juniper Street, Philadelphia, and who is now employed at 1331 Spruce Street, Philadelphia, five thousand dollars ($5,000). To any other household servants at 1331 Spruce Street who may have been in my service for two years or more and are still in my service at the time of my death, I give and bequeath the amount of two years' wages on the scale in force at the time of my death.

Fifth: I give and bequeath to the club known as "The Rabbit" near Church Lane and Belmont Avenue, Philadelphia, ten thousand dollars ($10,000) to be used as they may see fit. I also give and bequeath to "The Rabbit" all of the so-called "Certificates of Equitable Property Interest in The Rabbit" which may be held by me at the time of my death.

Sixth: I give and bequeath to The University of Chicago, Chicago, Illinois, to be used for the benefit of the Journal of Geology, of which I have had the honor of being one of the editors and associate editors for many years fifty thousand dollars ($50,000).

Seventh: I give and bequeath to the Economic Geology Publishing Company (incorporated under the laws of the District of Columbia) to be used for the benefit of the Journal known as "Economic Geology" twenty–five thousand dollars ($25,000).

Eighth: I give and bequeath all the rest, remainder and residue of my estate, real or personal, of which I shall be seized or possessed at the time of my death, in two equal parts, one of these parts to the American Philosophical Society held at Philadelphia for Promoting Useful Knowledge, and the other of these parts to the Geological Society of America (incorporated under the laws of the State of New York). Both of these gifts shall be considered endowment funds the income of which only to be used and the capital to be properly invested.

Ninth: I nominate, constitute and appoint my brother, Spencer Penrose, John Stokes Adams and The Pennsylvania Company for Insurance on Lives and Granting Annuities to be the executors of this my last will and testament.

Tenth: I order and direct my executors herein named to pay all inheritance taxes on my estate and on all legacies contained herein out of the principal of my estate so that all such taxes may and shall be charged out of the principal of my residual estate.

Lastly, I hereby revoke and make void any and all other wills by me at any time heretofore made.

IN WITNESS WHEREOF I have hereunto set my hand and seal this twelfth day of June Anno Domini one thousand nine hundred and thirty.

                          Richard A. F. Penrose, Jr. (Seal)

Signed, sealed, published and declared by the testator above named as and for his last will and testament in the presence of us who at his request and in his presence and in the presence of each other have hereunto subscribed our names in witness thereto.

                                        Wm. S. Andes
                                        Edward C. Lukens

Through Article 7 of the Penrose will, the Economic Geology Publishing Company received $25,000 as a direct bequest. This was paid promptly by the executors, though later a federal inheritance tax was levied and paid out of the estate. This tax was assessed on the grounds that the publishing company was owned by a group of geologist-stockholders, hence could not be considered a nonprofit, tax-exempt corporation for inheritance purposes (even though the stockholders never received dividends).

The Society of Economic Geologists, always separate from the publishing company, received no bequest, although Penrose had taken the lead in founding it. Soon after the will was probated, however, SEG requested financial assistance from the Penrose executors on the grounds that Penrose had promised to pay the costs of incorporating the Society and to start a bulletin of its own. In a compelling and detailed letter to the executors, SEG President L. C. Graton showed that the commitments had been incurred earlier but had not been billed to Penrose before his death. This claim led to strong feelings, both pro and con, in some quarters and somewhat delayed the final estate settlement, but it was eventually allowed by the executors and the court. The Society of Economic Geologists received a total of $28,521.82 from the estate.

The simplicity of Penrose's will arose directly from his experiences in settling family estates. In his letter of October 16, 1925, to his brother Spencer about the estate of brother Charles, he wrote:

All these complications about the division of estates seems to me to show the absurdity of anyone trying to influence posterity by making rigid trusts. If the trusts are made at all, they should be more or less elastic so that they can be adapted to the changes in conditions as time passes. The results of most trusts always seem to me to work a hardship and inconvenience on the descendants of the persons who create them.

In order not to add further to this complication at the time of my death, I have made a clause in my will leaving all my right, title and interest in the estates of our father and mother to you personally. This, I hope, will avoid further difficulties of the kind we are having now. [Fairbanks and Berkey, 1952, p. 650]

This letter shows that Penrose had drawn a will as early as 1925. Though he may not have then decided on principal beneficiaries or may have changed them later, the wording concerning the family fortune survived and became part of his final will of June 12, 1930. At the time of his death, Richard's share of the family trust funds plus his half interest in the family home went to his one remaining brother, Spencer. As last heir, and again acting under the trust terms, Spencer terminated it immediately and distributed the proceeds to himself and to the children of his brother Charles.

The news that GSA was indeed named to inherit half of the residue of the estate (Article 8) was enough to make GSA officers and members happy, but it was not until later when the true size of the inheritance was unfolded that they began to realize what had happened to the Society and to assess the responsibilities and opportunities that lay ahead. The size and makeup of the estate surprised everyone except, perhaps, his confidential secretary and his banker. Berkey and the few others in GSA who knew something about Penrose's plans would have perhaps guessed at a total value of something like $1 million; instead, its value as of date of death exceeded $10 million! The fortune had probably approached twice this figure a few years earlier, for it must have shrunk significantly with the market crash of 1929 and the onset of the great depression.

The makeup of the estate was as surprising as its size. Gone were the Utah Copper and other mining stocks that were the basis of the fortune; these had been sold, no doubt at near-top prices, during the booming twenties. In their place, the portfolio reflects a cautious and wise investor who foresaw the coming worldwide depression and who was determined to weather it.

Amazing to all who learned the facts, more than $3 million of the $10 million total was in cash! Most of this cash was on deposit with Penrose's Philadelphia bankers, Drexel and Company, but there were also large deposits in London, New York, and Denver banks. Aside from cash, one of the larger items in the investment portfolio, more than $600,000, was in U.S. Liberty Bonds. Nearly all the remainder was in a wide variety of state, municipal, and school district bonds, most of them issued by Philadelphia or other Pennsylvania entities.

This emphasis on tax-free "municipals" evidently reflected Penrose's desire to lessen his own federal income taxes. It did not, however, lessen a large estate tax by the state, and it soon called for a heavy turnover of the portfolio by the principal heirs. Both of them were tax-exempt organizations, hence tax-free securities were far less desirable than other investments that paid higher income.

Though the final settlement did not take place until 1934, virtually all of the estate was distributed only a year after Penrose's death. In part, the settlement was made so speedily because of the cooperation of Spencer Penrose. As one of the executors, he dissuaded certain family members not mentioned in the will from filing claims against the estate. Both of the principal heirs chose to take distribution in kind; they each took title on July 14, 1932. Much of the cash was needed during the settlement, so that the Societies' shares were taken partly in cash and partly by splitting the investment portfolio.

By the time of distribution, the estate had shrunk considerably. Further depreciation in security prices had erased about $1 million during the first year; state estate taxes, executor fees, payment of direct bequests, and other costs required more than another $1 million. Partly offset by continuing investment income during the settlement period, the final result was that the Geological Society of America and the American Philosophical Society each received a total of $3,884,684.42 (from Council Minutes report of Beatrice Carr, Treasurer, December 10, 1934). With acquiescence of the American Philosophical Society, GSA also received Penrose's library and most of his office furniture.

The story of how GSA adjusted to its sudden change from mere existence to wealth and of how it has managed its fortune in all the years since Penrose's death is told in two later chapters. The story of how it has used the income from the bequest in the advancement of geology, however, requires most of the remaining chapters of this volume. The Society's overall objectives are not much changed from those envisioned by the founders in 1888, but specific activities and accomplishments, and most particularly their scale, are all directly related to the income that still flows from the beneficence of one far–sighted and generous individual.

Many Were Called, But Two Were Chosen

Why did Penrose choose the American Philosophical Society and the Geological Society of America as his principal heirs? This question has intrigued many minds. There is no easy answer, but there is ample evidence that the choices were made only after long, careful, and thorough investigation of many candidates.

The decision to find heirs outside his family was an easy one for Penrose. He was a bachelor, and all his few remaining close relatives were amply provided for either through inheritance or their own efforts. Having eliminated relatives from serious consideration, he must have decided early to leave his wealth to science in some form rather than to any of the hundreds of other causes, worthy and unworthy, that are chosen by disposers of large fortunes.

At least as early as 1918, a dozen years before he wrote his final will, Penrose was almost obsessed with the idea of establishing some sort of Institute of Paleontology (or of Historical Geology or of Paleontology and Geology) in Philadelphia. This was partly because of sentimental attachment to the city of his birth, his family's seat for more than two centuries, and partly because he hoped to revive Philadelphia's place in the geologic world that it had held many years previously. Most of his efforts revolved around his old friend John M. Clarke, the famous paleontologist and Director of the New York State Museum. Clarke would have been more than happy to head up such an institution in Albany, but he successfully resisted the idea of moving to Philadelphia. His death in 1925 effectively ended the dream of making Philadelphia a geologic center. It coincided with Penrose's disillusionment, first with the University of Pennsylvania and then with the Academy of Natural Sciences of Philadelphia as hosts for his proposed institute. In 1922 he had accepted the presidency of the Academy with the single-minded purpose of guiding that body toward a strong commitment to the earth sciences. He retained the presidency for 4 years, devoting much of his time and energy to the Academy's affairs. Within 2 years, however, he had found such resistance to his dream of pushing the Academy into emphasis on a single discipline over all others that he abandoned the fight. His last two years as president were devoted to reorganizing the Academy and redefining its objectives. His resignation as president in 1926 marked the end of his thoughts of making the Academy his principal heir.

In a way, Penrose was repeating history when he tried to persuade the Academy to develop a deeper interest in geology. During the early stages of GSA's formation, a committee was appointed at the 1883 AAAS meeting in Minneapolis to "confer with the Mineralogical and Geological Section of the Philadelphia Academy of Natural Sciences. For various reasons, no meeting was called to discuss the subject at Philadelphia" (Fairchild, 1932, p. 70). There is no record as to the specific objectives of the proposed conference, but we can guess that the geologists in Minneapolis hoped for support in their breakaway from the AAAS.

Because it represented his own first love in geology and because he had more professional friends there than elsewhere, he must certainly have thought of endowing the Society of Economic Geologists (SEG). He helped found the Society of 1920, as well as its companion, the Economic Geology Publishing Co., to which he left a sizable bequest. As SEG's first president and donor of its Penrose Medal, he evidently believed in SEG and had high hopes for it. He always thought of it as a relatively small, elite organization that was highly selective in its membership, devoted to a small segment of the geologic spectrum—economic geology—and with emphasis within that segment on a single specialty, the application of geology to the search for ore deposits. (But when he endowed what was to become the Penrose Medal of the Society of Economic Geologists, he said explicitly that it was to be awarded for accomplishments in pure—not applied—geology!) His personal concepts of what the Society of Economic Geologists should be probably diverted him from any incipient thoughts that it might need a sizeable fortune to ensure its future.

The Wistar Institute of Anatomy and Biology, on whose board he served from 1915 to 1931, undoubtedly crossed his mind as a potential beneficiary. In clarifying his own plans, he had studied the terms of the Wistar endowment because he felt it was representative of the few endowments that had achieved their donors' purposes over many years. Also, he had offered the Institute the family summer home at Seal Harbor, Maine, for use as a scientific retreat, but the offer had been declined because the Institute foresaw no funds for upkeep of the place. Whether Penrose felt rebuffed by the refusal or whether the Institute's interest in anatomical and medical science was not his own first interest, there is no evidence that he ever thought further of making the Institute an heir.

The news that Penrose was seeking an heir must have seeped out through the scientific grapevine, for during the 1920s he received an increasing number of suggestions, as well as outright pressures, for financial help. As always, he replied courteously to all such requests and generally doled out innumerable gifts, ranging from a few to several thousand dollars, to causes that to him seemed worthy. Many requests he rejected out of hand either because they did not appeal to him or because the sponsors were too importunate. In addition to consideration of suggestions and requests from others, he must have considered all the usual forms of educational endowments—of buildings, of departments, of professional chairs, of new or existing institutes, and so on. Relatively early in his decision-making process he seems to have veered away from giving strong support to universities, largely on the grounds that objectives and methods change with changes in administrative and faculty personalities, so that a donor can feel no assurance that his gift will continue to be used as he intended.

In rejecting his desires that it revise geology as one of its major activities, the Academy of Natural Sciences of Philadelphia had unwittingly removed itself from an almost certain place as a Penrose heir. The pressure from John M. Clarke in Albany and Charles Schuchert at Yale to provide continuing support for their special fields of interest, stratigraphy and invertebrate paleontology, was particularly strong and persistent. They seem to have come closer to success than any others. Whatever his reasons, however,

Penrose finally decided against major support of any specialistic research or publication effort. On October 19, 1929, he wrote to Schuchert:

I am gradually coming to the conclusion that a national organization like the Geological Society of America, which is not bound to any particular institution but is on terms of friendship and good will with all of them, might be the best source through which to distribute the funds of a paleontologic or other geologic endowment. The very fact that the Society is nationaly in character gives it from time to time a broad knowledge of the distribution of talent in the branches of geology which it represents and hence would permit it intelligently to use funds to assist where for the time being certain researches are being pre-eminently pursued.

This broadening trend in his thinking eventually crystallized into a decision to divide his wealth between the Geological Society of America, as serving the whole of geologic science, and the American Philosophical Society, serving all useful human knowledge. No doubt his choice of the American Philosophical Society as one of his heirs was based largely on other reasons, but family loyalties and sentiment surely played a part. Penrose, himself, as well as his father and one of his brothers, had been honored by election to the Society; so too had numerous earlier relatives, some of them dating back to Revolutionary days.

Although we cannot know more about when and how he decided to help the American Philosophical Society, the gradual swing toward GSA was signaled several years in advance. Penrose was elected to GSA in 1889, less than a year after its founding, and he was number 151 on the Society rolls. Despite this early start, he took comparatively little active interest in Society activities, even when elected to the Council for 1914–1916. As councilor and Vice-President in 1919, as a member of the Finance Committee from 1924 through 1929, and culminating as President in 1930, his interest in the Society and its management grew apace. He endowed GSA's Penrose Medal during this period. He was deeply interested in the financial management and instituted important changes in the responsibilities of the Treasurer's office. He sympathized with the burdens of the undernourished and overworked headquarters staff, which then consisted of Secretary Berkey and Assistant Secretary Miriam F. Howells, both part-time employees and only one of them (Howells) paid any salary at all. He sent numerous checks of a few hundred dollars each to lighten their loads or at least to make them more tolerable. The most compelling clue to his intentions, however, lay in his efforts to reorganize the basic structure of the Society. This led directly to legal incorporation of GSA in 1929. While his intentions were perhaps not apparent at the time, this incorporation was considered by him to be a necessary prelude to the Society's legal ability to receive or handle trust funds.

During and after his presidential year, Penrose gave more and more evidence that he had made up his mind and intended to provide some sort of lasting support for the Society. He was particularly distressed that the Society's operations had to be carried on in the corner of a Columbia University professor's office and instructed Berkey to seek a separate and more permanent home. He told Berkey that he had provided for a headquarters building in his will and gave hint after hint that he recognized the need for an endowment fund to cover other purposes.

In deciding on the Geological Society of America as one of his heirs, Penrose was unquestionably guided by the highest scientific motives. It seems entirely possible, however, that there was another factor at work, one that is never mentioned but that deserves a few words here. This was the personal influence of Secretary Charles P. Berkey on Penrose's thoughts. Whether or not Berkey had any say in the nomination of Penrose as the Society's President will never be known, but this honor could not have come at a better time to sway the thinking of a man who was particularly sensitive to professional honors and to recognition by his peers—and who was at that moment approaching final decisions as to disposition of his fortune. Aside from this, Berkey and Penrose had been good friends, with much mutual respect, for some years before the 1930 presidency. They became even closer during that period for Penrose characteristically immersed himself in the welfare of the Society during his entire term of office.

In his letter of September 28, 1966, to then Secretary Raymond C. Becker concerning the proposed move of headquarters away from New York, D. Foster Hewett wrote as follows:

It may help you and others, if I as one of the few still living, who were on the Council during those years, 1931–1935 when we faced many problems arising out of the Penrose Bequest, recalled some of those days.

One of the great problems we faced was the location and nature of the headquarters of the Society. In this, Dr. Berkey, who had been Secretary for many years, without remuneration it should be recalled, was the dominating voice. In 1931, the only paid employee was Mrs. Howells, who also was Dr. Berkey's secretary at Columbia. [Hewett was in error here; Mrs. Howells worked only on GSA business—EBE.] Let me emphasize that we all felt the Society owed a great debt to Dr. Berkey and in many matters, deferred to his wishes. I should add, also, that *Dr Berkey felt very strongly that he was largely responsible for the Penrose gift to the Society and frequently asserted this* [italics added].

Hewett's is the only clear statement I have found in the files as to the part Berkey may have played in the final decision (though Helen Fairbanks, who worked closely with Berkey for many years and collaborated with him in the Fairbanks and Berkey (1952) book on Penrose, fully corroborates Hewett's memory in a letter to me). Addi-

tional insight, however, is suggested by a brief look at Berkey's special skills.

Berkey was highly successful as a teacher and as an administrator, both as long-time head of the department of geology at Columbia and GSA's Secretary. He was also an outstanding engineering geologist with a worldwide reputation as the father of that specialty. In all these careers, which he managed to pursue concurrently, his outstanding characteristic was persuasiveness. A master of the language, whether spoken or written, he could make abstruse ideas seem simple and easily understood; he could also make simple statements seem almost like revelations. This skill, combined with keen judgment as to the receptivity of colleagues, students, or other audiences, brought him to greater stature than would his acknowledged accomplishments in geology alone.

Is it possible that this same kind of persuasive skill may have been applied to Penrose during one or another of the many lunches, dinners, and office conferences that the two friends had? Berkey himself gives a strong hint as to his technique in dealing with a sensitive potential contributor to the Society. Speaking of the period before Penrose became President, he said, "Upon assurance that our thin stream of resources did not warrant further drain for this purpose [support of the headquarters office], he thereafter gave the Secretary, for addition to equipment and to the paid service of the central office, a few hundred dollars whenever it appeared to be needed. We never made the mistake of asking for money. We never admitted that the funds were entirely gone. We never gave occasion, either, for embarrassment or suspicion by solicitation or complaint" (Fairbanks and Berkey, 1952, p. 749).

In summary, the American Philosophical Society and the Geological Society of America came to inherit the Penrose fortune, not by whim or chance, but by a long, painful, and thoughtful process of elimination. Many worthy causes, both specific and general, were considered and ultimately dropped for one reason or another. Sentiment, personalities, hard business sense, life-long dedication to science—but with a desire to do more for science than he had been able to accomplish personally—all these factors and more came into play in leading to the final decision. We, the Geological Society of America, can feel proud and fortunate that Penrose found us worthy of a great trust.

GSA's Co-Beneficiary, The American Philosophical Society[5]

From time to time, Councilors, Treasurers, Investment Committees, or other members with real or casual interest

---

5. Abstracted from recent Year Books of the American Philosophical Society and from GSA Annual Reports.

in the Society's financial health raise questions as to the American Philosophical Society's (APS) stewardship of its half of the Penrose fortune as compared to that of GSA. Most such questioners believe or assume that the APS has done far better in conserving and building its Penrose endowment than has GSA. These beliefs are based largely on truth but also in part on ignorance and on the fact that it is difficult to compare apples and oranges.

First, a brief characterization of the APS and its activities is in order. The American Philosophical Society held at Philadelphia for Promoting Useful Knowledge is far older and very different from the Geological Society of America. It is, in fact, different from any other learned society; no other, for instance, can list 15 signers of the Declaration of Independence and 13 presidents, from George Washington to Herbert Hoover, among its former members. Founded in 1743 by Benjamin Franklin, who guided it through good years and bad until his death in 1790, the Society has always followed the broad original definition of philosophy as the love of knowledge or what we now call "science."

Membership in the Philosophical Society is honorary and has been restricted to election of the elite in all branches of learning; its roster, aside from a significant number of foreign members, is not unlike that of the National Academy of Sciences, and indeed there are many duplications between the two lists. Even today, its total membership is only about 600, one-twentieth the present size of GSA. This is comparable to the size of GSA when its membership was restricted to elected Fellows, but it must be remembered that Philosophical Society members represent all walks of the sciences, social sciences, and the humanities, whereas GSA members represent only one. Selectivity, therefore, is of a different order of magnitude.

The Philosophical Society's assets at the end of 1979 totaled about $23,752,000, exclusive of its valuable real property (on Independence Square in Philadelphia), library, and works of art. Much of its wealth is in the form of endowments; of these, the Penrose endowment is the largest, amounting to about $15,428,000 in 1979. GSA's assets totaled about $8,000,000, also exclusive of real property. About $6,600,000 of this total was in the Penrose endowment fund. None of the figures just cited are strictly comparable; the Philosophical Society's "Penrose Fund" is no longer identical with the income from its Penrose bequest, and significant parts of GSA's Penrose money are represented in the Reserve Fund, in the headquarters building, and elsewhere. Nevertheless, in the 47 years since each society received about $4 million from Penrose, the Philosophical Society's share nearly quadrupled, while the Geological Society's share was less than doubled.

Investment management of both groups seems to have been soundly conservative. The investment portfolios of the two societies are remarkably similar, at least in recent years, except that the Philosophical Society has tended to maintain a higher proportion of equities to fixed income securities.

The largest single item among the Philosophical Society's many activities is maintenance of one of the more important libraries in the United States, with priceless collections on Benjamin Franklin and Charles Darwin, on the history of science, of publications of other learned societies, and on a handful of other specialties, together with manuscripts, portraits, and statuary that record much of the political, industrial, and scientific growth of America.

The Philosophical Society's publication program consists primarily of two periodicals (*Transactions,* begun in 1769, and the *Proceedings,* 1838), the Memoir series (1935), and the *Year Book* (1937), plus occasional separate book publications. Despite the age, importance, and prestige of these publications, the program in recent years has been modest as compared with that of GSA, whether considered in terms of cost, of edition sizes, or as a proportion of total society expenditures.

Throughout its long life, the Philosophical Society has been dedicated to encouragement of research. This encouragement took modest form shortly after receipt of the Penrose bequest when it was decided that virtually all of the Penrose Fund income would be devoted to grants-in-aid of research. Supplemented by the income from one other large unrestricted endowment and several smaller restricted ones, the society now gives several hundred research grants each year with a direct expenditure of fully a quarter of its annual budget. Addition of administrative and publication expenses that relate directly to the research grants program means that support of research accounts for at least one-half of the society's yearly outgo.

Given similar investment programs, with generally comparable results, the two societies have diverged in the way they have used their Penrose gifts.

First, the Philosophical Society has interpreted the terms of the bequest more strictly than the GSA. It has chosen to use only the income from the Penrose Fund and has consistently reinvested all realized capital gains, thus adding to the principal fund. GSA, on the other hand, has for some years felt free to consider realized capital gains as part of spendable income, so long as the original corpus of the Penrose bequest is not invaded. This difference alone explains the differences in growth of the two funds. Even though the two societies may well have received and spent comparable total investment incomes through the years, the growths of their retained capital funds have not been at all comparable. They will probably diverge even farther in future years as the larger fund should produce larger income and capital gains. How much of the difference in money management has depended on advice of different attorneys or financial advisors and how much on philosophical differences would be difficult to assess.

Second, the Philosophical Society has chosen to use

much of its Penrose income for direct support of research, whereas GSA devotes its Penrose income to many purposes. Not only does GSA support a modest research grants program but also a large and expensive publications program and many services to many thousand members. It even called on the Penrose Fund to finance a fine headquarters building of its own, whereas no part of the Philosophical Society's Penrose bequest went to buildings. A headquarters building for GSA was something that the benefactor had much in his mind in the several years before his death, and the $1 million investment in land and buildings should really be carried as part of the present value of the Penrose Fund.

Third, the two societies have different living scales. The Philosophical Society's total expenditures for 1979, for example, were about $1,507,000, significantly lower than GSA's $1,901,876.

Memorials to R.A.F. Penrose, Jr.

The Geological Society of America itself is in many ways a living memorial to Penrose. Income from his gift is used to support at least partially virtually every Society activity.

Physical memorials to Penrose, identified as such, are few. They are: the Penrose Medal, established by Penrose as the Geological Society of America Medal but soon renamed in his honor by his colleagues; the magnificent biography, *Life and Letters of R.A.F. Penrose, Jr.,* by Helen R. Fairbanks and Charles P. Berkey (1952); Aitken's bronze bust of Penrose; and the Penrose Room and Library at Society headquarters. The Penrose Conferences are named in his honor. Though not labeled the Penrose Building, the GSA headquarters on Penrose Place in Boulder is a fitting memorial of the sort he would have liked. Less directly attributed memorials to Penrose and his farsightedness abound.

Shaler and Penrose Busts

Two fine bronze sculptures, busts of Nathaniel S. Shaler and Penrose, stand on display at GSA headquarters. They deserve a few words of explanation.

While studying at Harvard, Penrose was greatly influenced by Professor Shaler, one of the outstanding geologists and teachers of his generation. They became lifetime friends, and Penrose ever looked up to Shaler as his principal scientific mentor and advisor. After Shaler's death in 1906, Penrose sought some kind of lasting tribute to Shaler's memory. Penrose had strong aversions to research or scholarship funds that are so commonly established as memorials to departed scholars. Instead, he tended to favor statuary, fountains, or medals, which he felt provided more

lasting tributes and could serve as inspiration to coming generations.

Having successfully fended off pleas from Professor William Morris Davis and others (though not without some acrimonious exchanges) for money to memorialize Shaler in other ways, Penrose followed his own bent and commissioned the noted sculptor, Robert Aitken, to make a bust of Shaler. The original was presented to Harvard University, where it was displayed in the Faculty Room. Later Penrose asked Aitken to produce three replicas in bronze before the mold was destroyed. These went to the Kentucky State Capitol, to the Harvard Club of New York, and to Penrose's office in Philadelphia. This last copy is the one now owned by GSA.

Not long after Penrose's death, the Society used a small part of his gift to memorialize Penrose in the way he would have most wanted. Aitken was commissioned for the task. The resulting bust was unveiled during the New York Annual Meeting banquet in December 1935. Berkey's address at the unveiling is one of the more moving tributes to Penrose extant (Berkey, 1936). Today the bust stands in a recess outside the conference room at GSA headquarters, where Penrose can keep an eye on the Society's use of his fortune. The Shaler bust is in a similar recess on a lower floor. (Taken largely from Fairbanks and Berkey, 1952, p. 104–105, who quote an unpublished memorandum by Penrose dated August 20, 1928.)

The Bas-Relief Problem

In addition to the Aitken bust and the Penrose biography, a third memorial to the Society's benefactor was proposed not long after his death. Various forms of memorial were envisioned and discussed by the Council and by a special committee headed by Professor Alfred C. Lane. From these emerged a somewhat nebulous plan for a bronze plaque showing a portrait of Penrose in bas-relief, small replicas were to be made available for sale to the membership. Five artists were commissioned at nominal fees to sketch conceptual designs using photographs for models. At least two of them produced plaster casts or clay models as well as sketches. Opinions differed as to the resultant designs and even to the desirability of the project. Spencer Penrose further confused the issue by urging depiction of his brother as a younger man in the prime of life, whereas some others wished Penrose shown in his later years when they had known him best. Unknown to the rank and file, internal dissension smoldered for years. It was finally settled (?) in 1940 by discharge of the committee and "permanent" tabling of the project but not without wrankled feelings, a threatened lawsuit, and other unpleasantness.

The story of the bas-relief problem is scattered through the Council Minutes from those of 1932 to 1948. All the earlier bits were assembled by the Society's legal counsel, Allen R. Memhard, in a report to Council dated September 4, 1940. It was hoped that this report, in which Memhard expressed the opinion that none of the contestants had valid claims against the Society, had effectively closed the matter (Council Minutes for 1940, p. 265–289). This hope was short-lived. In October 1946 Professor Lane, chairman of the special committee and always the chief proponent for a plaque, told the Council that he had personally authorized one of the artists, L. L. Leach, to execute the bas-relief in bronze and had paid for it from his own funds. He sought the Council's permission to donate the plaque to the University of Chicago. Permission was, of course, quickly granted, but the accompanying sigh of relief was cut short when only a month later Lane told the Council that it was mistaken—he wanted the Society to have the sculpture after a brief loan to Chicago.

The matter was tabled again, but at long last, the 1948 Council refused Lane's gift on the grounds that acceptance would be a reversal of earlier Council decisions. (This was a far cry from the Council's more usual practice of reveling in its complete freedom to reverse actions of previous Councils.) The refusal coincided approximately with Professor Lane's death, and the two events did, indeed, close the bas-relief problem, except for one brief footnote: in 1978 the then members of the University of Chicago geology faculty wrote to me that they could find no records or other evidence that such a plaque as I described to them had ever been on display at Chicago or been in the department's possession! Perhaps some future historian will be able to shed more light on the mystery of the Penrose plaque.

Adjustment to Sudden Wealth

Other than the founding of the Society in 1888, the acquisition of the Penrose fortune was surely the most important single event in GSA's entire history. In many ways it was even more challenging to the leaders than was the birth of the new Society. The founding fathers had had years to perfect their plans to take over a group of geologists already organized, if loosely, within the American Association for the Advancement of Science and guide it toward advancement of the science of geology. Those in charge in 1931 and 1932 were faced virtually overnight with transition from mere existence to almost opulent wealth with responsibilities theretofore undreamed of for preserving a fortune and for using its proceeds wisely to benefit the science—in perpetuity. Fortunately, there were people equal to the task, not only to plan the sudden transition but to follow through on the plan for all the years since.

State of the Society in 1931

Consider for a moment the state of GSA in early 1931 just before the news of Penrose's death and his magnificent gift. The Society was made up of about 600 geologists, most of whom were approaching middle age when they were elected to fellowship on the basis of their scientific stature, not their business acumen. Its officers and councilors, too, were chosen largely for their standing in the profession, not for management abilities. The two chief activities were (1) the holding of annual meetings for camaraderie and the exchange of geologic information, and (2) publication of a quarterly *Bulletin*. Headquarters was a single room lent to the Society by Columbia University. The staff consisted of part-time Secretary Charles Peter Berkey, who also had more than full-time jobs in teaching and consulting; part-time Treasurer Edward B. Mathews, a Johns Hopkins professor in Baltimore; and part-time Editor Joseph Stanley-Brown, a financier who did his editing for GSA as an avocation. The Secretary and Treasurer received small honoraria, much of which went to pay the miniscule salaries of their part-time secretarial assistants. This was the GSA staff.

Financially, the pre-Penrose Society was more than solvent. In 1930, the last full year before news of Penrose's death and his gift, receipts and disbursements were in balance at about $12,000. Receipts were from dues, subscriptions to the *Bulletin*, and a small income from a $50,000 bond portfolio. Two-thirds of the outgo was for publication of the *Bulletin;* the rest went to office and miscellaneous expenses. There were no travel expenses for officers, Council, or committees; annual meeting costs were borne by the local hosts (except for the income provided by a $1.00 registration fee instituted in 1930).

In none of these conditions of genteel poverty was GSA different from any other comparatively small society of the period composed of individual members and led by volunteer officers.

Planning for Transition

Almost as soon as word of Penrose's gift reached the Society, the officers, the Council, and the staff swung into action and began to make plans for acceptance of the inheritance and for its wise use. Some actions were taken at once, but development of well-considered, long-range plans was wisely postponed for study by special committees appointed by the Council.

Immediate Actions

Among the immediate actions was the engagement of Allen R. Memhard as legal counsel. He was to guide the Society through the settlement of the Penrose estate and transfer of GSA's new fortune. For many years thereafter his legal opinions were to have a profound effect on the Society's investment policies and their results, as well as on innumerable other facets of its new life.

Other immediate actions included polling the membership for suggestions as to how the Society could best meet its new responsibilities, and a move to acquire a headquarters building. Knowing the strength of Penrose's beliefs concerning GSA's need for a home of its own, Berkey took immediate steps to arrange with Columbia University for the indefinite loan of one of its residence buildings in uptown New York City. The house at 419 W. 117th Street on the edge of the campus was destined to be the Society's home for 30 years; it is further described and pictured in another chapter. Necessary alterations to the building were begun at once, headquarters moved there from its room in the geology department in April 1932, and the Council first met in the new home two months later.

One other action that took place immediately is worth noting because it, too, had far-reaching effects on the Society. Joseph Stanley-Brown resigned the editorship, loading the duties of that office on the already overloaded shoulders of Secretary Berkey. Thenceforth Stanley-Brown devoted all of the time and energy he had available for GSA to the chairmanship of the Finance Committee. His nearly 50 years of devoted service to GSA as Editor and as financial advisor are described elsewhere (see Chapter 10).

Still another of the early decisions was the donation of $100,000 to support of the XVI International Geological Congress to be held in the United States in 1933. This gift was the first of many to other organizations in subsidy or support of the science. The International Congress matter had been under discussion in the geologic community for some time before Penrose's death, and several nebulous plans for financing it had been advanced, both in and out of the Society. With the news of the Penrose endowment, those plans were scrapped; instead, the Council in April 1932 earmarked for that purpose the income from Penrose's investments that would accrue between the time of his death and the estate settlement.

One more significant decision was made very early. Secretary Berkey was given a half-time salary of $6,000 in place of the customary honorarium, and travel expenses to meetings by the President, Secretary, and Treasurer (but not of other Councilors) were authorized for the first time. Berkey forewent his salary after a short while and reverted

to an honorarium, but the precedent for paid staffs and travel in Society management had been established.

All through the transition period, the Council and its special committees evidently had troubles in adjusting to the realities of sudden wealth. Minor details of Society management still followed the old, cumbersome methods of group description, extended discussion, and final decision. Budgets and investment recommendations were subjected to careful scrutiny, and requests for modest salary increases or of the tiny allowances for miscellaneous expenses were, more often than not, cut or put over for future meetings. But this changed gradually as Councilors began to realize the enormity of their own responsibilities and of those of a paid staff that was totally inadequate in terms of its numbers and its remuneration but not in its dedication.

Advisory Committees

Many new committees and appointments were made in the wake of the Penrose bequest. Two of the committees in their advice to the Council had lasting and significant effects on the entire future course of the Society, and their reports must rank as among the most important in Society archives. These were the Advisory Finance Committee, further discussed in Chapter 8, and the Advisory Committee on Policies and Projects, described here. The work of a third group, the Special Committee on Revision of the Bylaws, was of equal importance to the welfare of the Society, but its duty was merely to codify the decisions reached by the Council as a result of the Policies and Projects Committee report. Its report, signed by D. Foster Hewett, Nevin B. Fenneman, and Charles P. Berkey, was presented to the Council on November 5, 1932, and the new Bylaws were approved by the Council and the membership with no essential changes at the annual meeting in December 1932 (*Bulletin*, v. 44, p. 109). These Bylaws are far more detailed in defining duties of officers, and committees, and of all the Society's activities than any Bylaws before or since. They amounted to a superb set of Council Rules, Policies, and Procedures such as have been published separately and intermittently in recent years.

Advisory Committee on Policies and Projects. This committee was appointed by the Council at the annual meeting at Tulsa in November 1931; its report was submitted in April 1932. The committee, obviously wisely chosen, consisted of C. K. Leith, Chairman; F. D. Adams; Isaiah Bowman; R. T. Chamberlin; M. M. Leighton; Waldemar Lindgren; and J. B. Reeside, Jr.

The report was based on the collective wisdom of the committee members, on about 150 letters of advice from GSA members, and on personal conferences with many Fellows and with the heads of various research institutions. Chief elements of the recommendations are summarized here, but the committee's logic and reasoning that led to the recommendations are condensed or deleted. The full report appears in the Council Minutes for 1932 (p. 179–203).

1. *Location of Headquarters.* Accept Columbia University's offer of a five-story house adjoining the campus; postpone consideration of building a permanent home. (The membership had expressed widespread objections to establishment of a permanent home regardless of location and to development of an extensive headquarters staff.)

2. *Organization*
    a. Concentrate all administrative service under one roof, with Secretary Berkey to be in charge, with pay for half his time.
    b. Employ the following full-time staff: an Assistant Secretary, one secretarial assistant (additional to Mrs. Howells), an Assistant Treasurer, and a stenographer. (This staff, with upkeep on the headquarters building, was estimated to cost about $18,000, or 13% of anticipated income.)
    c. Appoint three administrative committees on (1) projects, (2) publications, and (3) finance. These committees are to report to Council, and their chairmen should not be voting or ex officio Counsilors.
    d. Eliminate elective office of Editor; all editing to be handled by Secretary's office, largely by an Assistant Secretary. (This recommendation did not take full effect until 1934 when Henry R. Aldrich became Assistant Secretary and assumed the editorial duties.)
    e. Make the secretaryship appointive, rather than elective, and make him an ex officio (voting) member of Council. (Not until 1962 did the Secretary lose his place on the Council.)
    f. Define the status of the Treasurer only after the Finance Committee has reported but consider centering financial activities at headquarters, making treasurership appointive rather than elective, and not adding him as an ex officio member of Council. (This recommendation was not accepted; he remains an elected officer and Councilor to present day.)
    g. Appoint a three-man Executive Committee to act between Council meetings.

3. *Library.* Preserve the Penrose library as a dignified memorial to him, but do not attempt to build up a great library around the Penrose nucleus; on the other hand, do not disperse Penrose's volumes as has been urged by some.

4. *Travel Expenses.* Restrict travel expenses to payment of rail and pullman fare and meals en route, but not board and room at meeting place, for officers, necessary assistants, Councilors, and administrative committees.

5. *Aid to Local Committees.* Allot some uniform sum, say $500, to meet routine expenses of local committees in hosting Annual Meetings.

6. *Aid to Other Societies.* Subsidies to other societies and research organizations are not recommended, except under special circumstances and where GSA could retain control. (This recommendation was never fully adopted.)

7. *Aid to International Geological Congress and Other Scientific Meetings.* Except for the 1933 IGC, to which GSA has already committed $100,000, not recommended (and probably never again financially feasible).

8. *Remission of Dues.* Cancel dues of Fellows over 70 years old who have been in GSA for 30 years.

9. *Prizes (in addition to Penrose Medal).* Not recommended.

10. *Fellowships, Scholarships, and Professorships.* Not recommended, except rarely to help complete approved research projects.

11. *Publications.* Improve, expand, and distribute pubications more widely:
    a. Improve *Bulletin* in content, format, and appearance.
    b. Establish a Memoir series (but beware of unloading of expensive volumes by other research organizations!).
    c. Establish a bibliography of the geologic literature of the world (in conjunction with U.S. Geological Survey which already publishes the *Bibliography of North American Geology*).
    d. Abstracts. Recommend against an attempt to publish abstract journal or to aid others already started or contemplated.
    e. Wider distribution of publications. Abandon present exchange system and establish very broad system of foreign and domestic depositories (estimated 427 in addition to the 600 GSA members) which will receive all publications free.
    f. Improvement of publication standards. Several recommendations involving Editor (Assistant Secretary) and Publication Committee leading to tighter selectivity of manuscripts, tighter editing, and guidance to authors; also recommends tighter control of quality and quantity of papers presented at annual meetings.
    g. Publication Committee. Establish a Publications Committee to set policies and standards.

12. *Research Projects and Policies*
    a. Give short-time allotments in financial aid to individuals or organizations in preparation of reports on noteworthy projects already far advanced but lacking means for completion or publication.
    b. In exceptional cases, give grants to experienced workers to start new research projects.
    c. Establish a small Projects Committee appointed by Council.

The Committee based its recommendations on the premise that GSA is primarily designed to maintain and improve standards of geologic research, to serve as a forum for expression of thought, and to afford adequate publication facilities. It felt strongly that the Society should not aim to mount or conduct research projects in its own right. It included a list of more than 50 research projects, large and small, that had been suggested by members.

The report closed with a reminder that its proposals would require changes in the Bylaws and with a summary estimate of the annual cost of the program. This totaled $160,000, about equal to the expected income from all sources.

With but few exceptions, the Council agreed with the Committee's recommendations and took the necessary actions. Some of these were immediate, others required more time. (Interestingly, an Executive Committee was established over the objections of the Treasurer, the Secretary, and the chairman of the Finance Committee—the three men who had until then had more to do with managing Society activities than any others.) Within 3 years after Penrose's bequest, however, the Society had made the transition from genteel poverty to wealth and had embarked on a course that would serve the science well for many years to come.

Not surprisingly, by mid-1933 and before the transition to wealth had even been completed, there were criticisms and complaints from the membership as to the high cost of administration. The Finance Committee, then consisting of Joseph Stanley-Brown, Chairman; Donnell Foster Hewett; William E. Wrather; and Edward B. Mathews, was asked in October 1933 to advise the Council. This it did at the November meeting. Despite the shortness of time, the committee studied the problem in depth. It concluded that administrative costs were not out of line and saw no way to reduce them very much and still permit the Society to do all it wished to do. It did revise the 1934 budget to some extent and recommended some changes in organization that seemed called for by experience of the last several years. The grassroots complaints that led to this special study of the Society's administration may have been the first in GSA's history, but they were by no means the last.

Part II

Society Makeup, Governance, and Management

Led by elected Councils and officers and served by
volunteer and paid staffs, the Society grows from 100 to
12,000 members; it adjusts from genteel poverty to
affluence, learns to use its money wisely, and finally settles
in its own home.

Membership

Society membership totaled 12,603 at the end of 1980. A chart indicating the growth record from the original group of about 100 founders in 1888 is shown later in this chapter. Before we look into the striking breaks in the curve and the reasons for them, let us examine the reasons why thousands of geologists have seen fit to join GSA and to maintain their memberships through the years.

Rewards of GSA Membership

Many geologists believe that election to membership and especially to fellowship in GSA marks peer recognition of their stature in the profession. These beliefs are based on fact, at least in part. Memberships and activities in the "right" societies are considered by most employers both in hiring and in granting promotions. It is interesting that at least one other society with very rigid requirements as to education and experience considers membership in GSA as a strong plus factor in admission of new members. This, even though any geologist with a bachelor's degree and a few dollars for dues is welcomed by GSA.

In former days, fellowship perquisites were substantial. Fellows were considered to have vested and exclusive rights to publish in the *Bulletin,* and only they or guests introduced by them could present papers at GSA meetings. For many years, too, research grants went only to mature Fellows. More recently these class distinctions have disappeared. Papers for either oral or printed publication are accepted from anyone, and most research grants go to beginners rather than to more mature geologists.

Aside from its prestige—and ego-building value—membership or fellowship in GSA means different things to different people. Some join the Society because of a deep-seated belief that it is part of a scientist's duty to support the group that supports and represents his own interests. Others join primarily to have regular and convenient access to the *Bulletin* or to *Geology*, or to get discount privileges on purchases of book and map publications. Still other geologists may join in hopes of priority treatment in invitations to Penrose Conferences, in the Society's employment service, or in acceptance of papers for oral or printed outlet.

Whatever their reasons, selfish or unselfish, geologists do join the Society in significantly large numbers. More important, most of those who join remain with the Society for life. Dropouts for other than death or disability have remained remarkably low through GSA's history.

Membership Classes

Today there are four membership classes (see accompanying table). One of these classes, that of Honorary Fellows (formerly called Correspondents), is strictly invitational and is composed of a small group of internationally respected leaders in the science, almost exclusively from outside North America. It is described in the section on Honors and Awards, Chapter 12, and is not further discussed here.

GSA MEMBERSHIP

Class	Class established	Totals Dec. 31, 1980
Honorary Fellows (Correspondents)	1888	40
Fellows	1888	3,229
Members	1948	7,500
Student Associates	1971	1,834
		12,603

According to the Bylaws (Art. 1, sec. 3) "Fellows shall be elected by the Council from the Members of the Society." In practice, election to Fellowship is considered to be an honor accorded to those who have achieved some stature in the profession. Recipients of the Penrose and Day medals become Fellows automatically if they are not already so classed; all others are nominated by their colleagues, considered by the Membership Committee, and elected by three-quarter vote of the Council.

The only requirements for Members are that they express a wish to join the Society and that they have a bachelor's degree in geology or related science, or equivalent training and experience. Student Associates must be full-time students (undergraduate or graduate) in geology or related science in a degree-granting institution.

Other Classes

The original Constitution established a class of members to be known as Patrons. Like the provision for Correspondents, this was designed to be an honorary class, to pay some tribute to those who had contributed largely to the Society's welfare in money or in other ways. Although there have been many such contributors through the years, most notably R.A.F. Penrose, Jr., and although the term "patron" has been applied informally to him and to several other people in acknowledgements of generosities, I find no records to indicate that any Patron was ever actually so designated. The classification was long since abandoned. A subclass of Honorary Life Member was carried on the Treasurer's books for some of the early years. Designation as a member of the class, with lifetime remission of dues, was intended to honor those who had rendered signal service, thus was a poorly defined parallel to the Patron class. Only four recipients are on record (Fairchild, 1932, p. 109): John J. Stevenson, 1891; W J McGee, 1893; Henry L. Fairchild, 1899; and Joseph Stanley-Brown, 1900.

The class of Honorary Life Member has been abandoned, although on rare occasions the Council has voted to reward outstanding service to the Society, such as volunteer work as Editor, by lifetime remission of dues (but without bestowal of a special title or other fanfare).

A final class of membership was a purely economic one—that of Life Member. The original Constitution provided that any duly elected Fellow could commute dues for life on payment of $100. Later raised to $150 (but still a bargain), this fee bought all the privileges of membership, including receipt of all publications, for the life of the foresighted buyer. Relatively few Fellows ever took advantage of the opportunity, but the plan did result in addition of significant capital to the Society coffers, particularly in the early days. Whether life memberships were ever actuarially sound, even when dues were only $10 per year, is debatable; they became less and less so as time moved on. In 1947, the privilege of buying life membership was extended to the new Member class as well as to Fellows, but it was restricted to those who had already paid annual dues regularly for 15 years. No new Life Members or Fellows were added to the rolls after 1947, and the scheme was rescinded in 1963 with adoption of the new Constitution and Bylaws. Those few who still hold life memberships have experienced remarkable returns on their original investments.

From 1947 onward, and in response to the ever-increasing costs of the Society's publications program, various attempts were made to limit the privileges of the Life Member class. Most of these aimed to allow the Life Member only those publications that were available free to other dues-paying Fellows and Members. Strangely, most such decisions seem to have been forgotten soon after they were made. Each new financial crisis with calls for retrenchment resulted in frenetic record searching by the staff to determine exactly what the privileges of Life Members were. The last such search was in 1978, when the President wrote apologetically to the few remaining Life Members that henceforth the publications sent free to them would be severely restricted.

The action to limit the sending of free publications to the various classes of dues—exempt Fellows may have been necessary for economic reasons. The method of doing it, however, and the way in which those affected were notified were not particularly well received. Whatever the reasons, the abrogation of Society promises made many years ago resulted in unhappiness and disaffection of some of the Society's older and hitherto most ardent supporters.

Still another membership class exists, primarily in the Society's membership and accounting records. Loosely defined, it consists of all those of whatever category who are exempted from paying dues. These include Honorary Fellows, medal recipients, long-time dues payers exempt for age, and certain disabled members. Under certain circumstances, enlisted men and women in the armed forces and Member spouses of GSA Members were partially or fully dues-exempt. As noted above, the Council has on rare occasions granted lifetime dues exemptions for signal service, but these have been rare.

The Society—Honorary or Democratic?

The Exclusive Fellowship, 1888-1947

In the very beginning GSA was truly democratic, but it soon began a trend to exclusivity that lasted for many years. Initially all applicants were welcomed to Fellowship so long as they were working in the field of geology or teaching that science. Any tendency to exclusivity extended only to students who were not yet earning their living as geologists and to nongeologists. Evidently the Fellows perceived themselves as an identifiable and respectable group of professional persons and believed that all were entitled to belong to the Society that was to represent them. Nongeologists were discouraged from applying for fellowship only by the deliberately high annual dues ($10) and presumably by lack of publicity about the new Society outside of the geologic fraternity. By the time of the second annual meeting in 1889, virtually all practicing geologists in the United States and some of those in Canada had been elected to Fellowship (see Chapter 2).

During the next half century (through 1947) the Society became ever more exclusive. The change was gradual, and few of the records even hint at any conscious change in requirements for Fellowship or any realization that the Society was becoming more selective.

The numbers of those elected to Fellowship each year and total membership are on record. No comparable records as to the total numbers of North American geologists exist, however, so there is no easy way to prove the assertion that GSA Fellows made up progressively smaller proportions of the profession. Such data only began to become available during and shortly after World War II. To accept the assertion that the Society was gradually becoming more exclusive, it is only necessary to note the low slope of the membership curve through 1946, which shows that total membership less than doubled during the 1910–1930 period when petroleum geology, only one of many specialties, was burgeoning and when the American Association of Petroleum Geologists, not organized until 1917, was already outstripping GSA in total membership.

The original Bylaws stated simply that "Fellows shall be persons who are engaged in geological work or in teaching geology." This definition of requirements survived with only minor changes until 1947, almost 60 years. It was still relatively unchanged in the Constitution and Bylaws operative in 1980 (Appendix B). Throughout the long period of 1888–1947, much more than half of the Society's life to date, each candidate for Fellowship who had survived preliminary screening was voted on by the entire Fellowship.

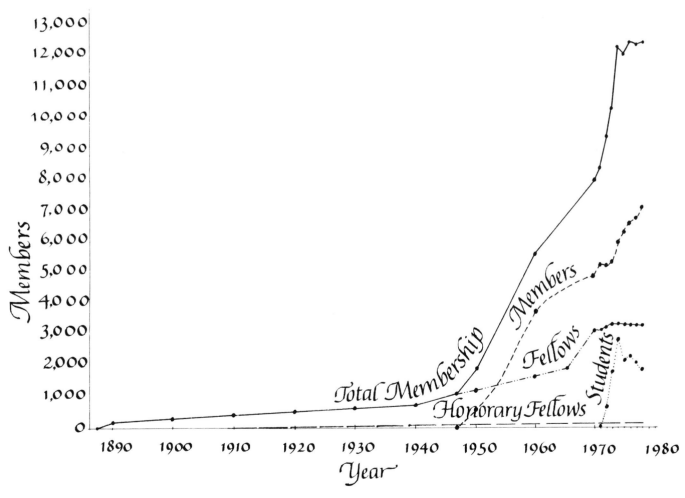

GSA MEMBERSHIP

Requirements changed from time to time, but in general, negative votes of about 10% of those voting resulted in disqualification. In reality, this meant that a candidate could be barred by less than 5% of the total Fellowship because less than 50% exercised their franchise. However, very few candidates ever failed at this stage of the election process.

Before candidates' names, biographical records, and summaries of achievement were submitted to the Fellowship, they were carefully screened first by informal or formal membership committees and then by the Council. At each stage in this long and arduous process, candidacies could be turned down or, perhaps more frequently, sent back to the sponsors for more information. Many nominations were denied on the grounds of prematurity, meaning in effect that the candidate had to do or publish more work to prove he was advancing the science of geology or perhaps only that he needed to grow a bit older chronologically.

Election to Fellowship came to be considered a high honor, to be attained only by invitation. Few candidates ever dared volunteer themselves for possible election. Instead, nearly all were nominated (with their permission) by friends and colleagues. This custom alone tended to lead toward exclusivity since some institutions were alert to seek out their more promising achievers and to arrange for their nominations and others hesitated to do so or could not because they were remote from geologic population centers.

In the absence of specific criteria for Fellowship and subject to the whims of constantly changing committees and Councils, it is easy to see why many good geologists never became Fellows, why the Society grew more and more slowly as compared to the overall geologic population, and why the Fellowship grew ever older and more elite. By 1944, for instance, the average age of Fellowship was 55 and the average age on election was 40, both higher than most professional societies and notably high for a science that depends so heavily on demanding and vigorous field work for most of its advances.

Coupled with the pressures and uncertainties described above, and perhaps more effective than any of them in promoting exclusivity, was gradually increasing emphasis on published products as the main criterion for admittance. This emphasis was never adopted as stated policy, but Council Minutes during the 1890s and the first decades of the present century reveal preoccupation with the quantity (and secondarily the quality) of a geologist's written word as proof of his eligibility. This equating of publication record and readiness for Fellowship is still extant, perhaps even more actively since the Society permitted two classes of membership. This is not unique to GSA—promotions and tenures in academe and advancements in government or private organizations tend to worship the same god.

This problem is discussed at greater length in later pages, but it must be recognized here as one of the outstanding factors in the development of a highly selective, elitist Society.

The Quasi-Democratic Society, 1948–

Abruptly, in 1948, the Society at last became a quasidemocracy. The Bylaws were changed to establish a class of Members to which anyone with a bachelor's degree in geology or related science and who was working or studying in the field could belong. Many years later, in 1971 democratization advanced another giant step when a class of Student Associates was established. This class, open to any bona fide student in geology, made most of the privileges of GSA membership available to the geologists of the future, and at reduced rates.

Neither the 1948 nor the 1971 changes in Society makeup meant that it had become a democracy in the true sense of the word. Rather, it became a two-class and later a three-class Society. The Fellowship class was retained, and until recently only Fellows could serve as officers, Councilors, or committee members. Members were permitted to vote in general elections but had no other voice. Student Associates, perhaps rightly, still have no voice in Society affairs.

In 1972, by a constitutional change, Members attained the right to hold office on equal footing with Fellows. In practice to date, however, this privilege is almost meaningless. Many Members, as distinct from Fellows, have served as officers of divisions and sections, and a few have been appointed to Society committees, but as of 1980 no Member had ever been elected to the Council or as an officer. Perhaps this is as it should be. The Society needs stature, maturity, judgment, and experience in its leadership, and though there are many obvious exceptions, these qualities are more likely to be found among the Fellows than among the generally younger Members.

Addition of a Member class to the Fellow class appears from the bare record to have been abrupt. Like almost every other major action taken by the Society, however, this one had many precursors. At least as early as 1925 the Council Minutes record a suggestion that there be two classes of membership. A committee was appointed to study the suggestion, but no further action was taken.

From 1925 onward voices were occasionally raised against the Society's exclusive policy. These came from within the Fellowship, the Council, and the officers. After receipt of the Penrose fortune in 1931, when the Society was suddenly able to expand its publications and other activities hence to have more perquisites to offer, many other voices were raised by outsiders. Policies were unchanged, however, until the 1940s when the yeast of liberalization began to work. The first evidence passed almost unnoticed—lists of nominees and of newly elected Fellows began to lengthen, strongly suggesting that the bars of selectivity were gradually lowering. More important was the experience of GSA and of other geologic groups in establishing the place of geology in nations engaged in total war (see Chapter 16). Immediately after World War II the tide turned violently, and in December 1947 the electorate adopted new Bylaws that opened the Society to a new class of Membership. The arguments used to persuade both Council and the Fellowship that democracy's time had come were admirably summarized by President Gilluly in one of the more persuasive and thoughtful annual presidential reports in the Society's history.

First, I believe the Society should remember its purpose: "the advancement of the Science of Geology in North America." Our policy should be framed with this objective and this objective only. In the structure of North American science geology is in the opinion of most of us a far too inconspicuous member. Furthermore, in the competition for research funds, whether from the present grants of the armed forces or from a future National Science Foundation, we must expect to suffer from our lack of weight. Our total resources available for research are a small fraction of those that should be available if geology is to keep pace with its sister sciences and with its obligations to society. Will the science be better promoted by an exclusive organization directed, let us flatter ourselves, by "contributors" to the science, or by a society that embraces as large a proportion of the geological fraternity as possible? How sure can any of us be of our appraisals of each other's contributions? Will a Society of 1200 Fellows carry as much weight in national councils related to Selective Service, to educational subsidies or to research grants as, let us say, one of 8000? Are more recruits to the science apt to be attracted by 2000 copies of our publications scattered over North America or by 10,000? The objective of the Society is *not* prestige for its Fellows but the advancement of the science. And who will say that the prestige of our society is higher than that of the Chemical Society with its 55,000 members, the Psychological Society with 10,000 or the American Medical Association with scores of thousands? Or of our sister society, the Geological Society of London, whose membership policy is far more liberal than ours? However that rhetorical question may be answered, there is a clear record during the War of which of these agencies carried weight in national councils. It was not The Geological Society of America! Today nearly every university seeks for geologic staff—essentially no graduate students completed their training between 1942 and 1946. The shortage is there in other fields as well but no science is worse off than ours. In my view the science of geology has been handicapped for many years to come vis-a-vis the other sciences and a major factor in this is the divisive tendency within the science—a tendency that was nurtured by our exclusive policy of a generation ago and is being scrutinized today by the same, to me, short-sighted policy. [Gilluly, 1947]

Gilluly's report to the membership in 1947 was foreshadowed by several years in another compelling but unpublished document. This is the April 14, 1944, draft report

of the Membership Committee to the Council (Council Minutes for 1947, p. 84–87). It was prepared by committee members A. I. Levorsen, T. M. Broderick, and James Gilluly, Chairman.

The Councils of 1947 and 1948 were among the most active and revolutionary of any in GSA's history. Not only did they take the steps to democratize the Society, but they initiated a major change in investment management (see Chapter 8) and separated geographic and technical groupings into sections and divisions (formerly all were called divisions). Among many other actions, they permitted election to Fellowship by a three-quarter vote of the Council rather than by vote of the entire Fellowship. (This method was still in use in 1980, but according to the Bylaws, Art. 1, sec. 1, Members are now elected directly by the Membership Committee with ratification by the Council after screening by the staff.)

Admission of a Member class in 1948 did not solve all of the Society's membership problems by any means. It was obviously a popular move, as can be seen in the immediate and continuing rise in total membership. For the first time since the 1890s GSA could rightfully claim that it represented a respectable percentage of North American geologists. But it was (and still is) a two-class Society. Careful distinctions were drawn between Fellows and Members. Fellows retained all management functions, they continued to receive virtually all publications free, and they paid higher dues. Most important and for the first time, accomplishments other than publication were permissible evidence of eligibility for election. Any one (or preferably more than one) of the following criteria, were to be considered: (1) research (represented primarily by publications), (2) administration of geologists, (3) training and teaching of geologists, and (4) other activities of benefit to the science or to the Society.

Members had no voice in Society management other than the privilege of voting, they received fewer free publications, and they paid lower dues than Fellows. They needed only a B.S. degree in geology or its equivalent to qualify. Most of these class distinctions have vanished through the years by various Bylaw changes, partly by attrition of Fellowship privileges and partly by additions to Membership privileges. The gap in relative prestige between Fellows and Members, both in and out of GSA circles, remains unchanged.

Problems of a multiclass society, and particularly those of criteria for elevation to the Fellowship class, have continued to haunt Society management ever since the Member class was initiated. So far as can be seen ahead, they will continue to haunt those in charge so long as GSA has more than a single class (and the experience of the first 50 years shows that even a one-class Society was not problem free).

Virtually every Membership Committee has wrestled

with the problem either in consideration of specific Fellowship nominations or in searches for better general guidelines and pleas for advice. Councils, too, have wrestled with the problem and spent untold hours in discussion or in study of complaints from the membership or calls from the committee for better guidelines. As early as 1949, an open forum was held at the El Paso Annual Meeting to discuss qualifications for Members and Fellows. More than one membership poll has been taken by mail. Predictably, perhaps, most polls have been inconclusive—those who have already attained Fellowship tend to want to retain the status quo, whereas Members want to drop the multi-class system or do not express strong opinions either way. Eventually, perhaps, the Society will evolve toward a truly democratic system, but this seems unlikely to happen soon.

Employment Distribution

For much of its life, GSA drew most of its members from among university teachers of geology and from both federal and state government agencies. Academic people were probably in the majority in the earlier years, but with gradual growth of all governments and with proliferation of state and provincial geological surveys, the proportion of government employees was greatly increased. Few firm statistics exist, and none would be very meaningful because of the widespread practice of part-time employment of teachers by governmental agencies.

As employment of geologists by industry grew, so too did GSA membership by company geologists and consultants. To almost everyone's surprise, a count in early 1979 showed that GSA membership was almost equally divided between academia, government, and industry. Excluding students and retired Members and with two-thirds of the total membership responding, 32.4% reported themselves as self-employed or in industry, 32.2% were teaching, and 26.6% were in government employ (*GSA News & Information,* March 1979, v. 1, no. 3, p. 33). By early 1981 the distribution had changed to about 46% industry, 31% teaching, and 24% government (*GSA News & Information,* March 1981, v. 3, no. 3, p. 38).

Geographic Scope

Recurrent for many years was the question as to admittance of non-North American geologists to Fellowship (and later to Membership). Foreign Correspondents (Honorary Fellows) had been elected since the earliest days, but these were all men who had reached the pinnacles of the science, their election was honorary, and their numbers were traditionally restricted. There have never been more than 60 living Correspondents. The question of admitting other non–North Americans to GSA has been considered and reconsidered, however, by many Councils. It was finally settled, probably for all time, when the new Constitution and Bylaws adopted in 1963 opened the doors to geologists from anywhere in the world.

The persistence of this little problem is all the more surprising when one learns that admittance of "foreigners" had begun prior to 1894. Residence in North America was deleted from the requirements for Fellowship in 1894 by a change in the Constitution (Fairchild, 1932, p. 143). There was already one South American Fellow, but he had been elected on the "fiction" that his residence in Brazil was temporary. The proposal to change the residence requirements, even though vigorously supported by outgoing President G. K. Gilbert and others, was defeated by the membership in 1893 and had to be resubmitted for another vote in 1894 when it passed. By 1924 a total of 11 foreign Fellows from Mexico, South America, Europe, New Zealand, and Australia had been elected.

Until 1963, the purpose of the Society was always stated in some words as "The promotion of the science of geology *in North America.*" This geographic restriction caused endless confusion not only as to who could be elected to membership but also as to the subject matter of papers to be published and even as to where Society-supported research might be done. In practice, the restrictions were more and more loosely interpreted as time went on, though not without many time-consuming debates and formal or informal searches of the records for precedents. Finally, in 1963 the Committee on Revision of the Bylaws described the Society's purpose simply as "the promotion of the science of geology." This was belated formal recognition of the facts that both geology itself and the membership of the Society were worldwide.

Women in GSA

Almost from its beginning, women geologists have played a small but important part in the Society. Though there had been a few women geologists earlier in the 19th century (Arnold, 1977; Arnold, L. B., 1979, dissertation in preparation), none of them took firm places in the profession until toward the end of the century.

In admitting women scientists to its ranks, GSA was far more liberal than the venerable Geological Society of London. GSA elected its first woman Fellow when the Society was only a year old. The Geological Society of London, on the other hand, was already 112 years old in 1919 when it elected 12 women. Part of the impetus even then was to replace men whose ranks had been depleted by World War I.

Mary Emilee Holmes, a paleontologist, became the first woman Fellow of GSA in 1889. She graduated in 1870 from Rockford College, then Rockford Female Seminary, where she studied music and taught Spencerian penmanship. From 1877 through 1885 she returned to her alma

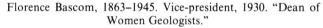

Florence Bascom, 1863–1945. Vice-president, 1930. "Dean of Women Geologists."

Mary Emilee Holmes, 1850–1906. First woman Fellow of GSA, 1889.

mater to teach natural science. In 1887 she received her Ph.D. degree in paleontology from the geology department of the University of Michigan.

Alexander Winchell, father of GSA, was chairman of her committee. University records show that hers was the first Ph.D. degree in geology to be awarded to a woman in America. Miss Holmes published but one paper, her dissertation (Holmes, 1887). Apparently she practiced in the profession for only a short time after receiving her degree. With a number of other women, she did appear on the program of the World's Congress on Geology held in Chicago, August 21–26, 1893, during the Columbian Exposition. There she presented a paper on "Methods of Teaching Geology." But Miss Holmes had evidently already lost her vocation for geology and heard a higher call for in 1890 she established a memorial school in her mother's name at Jackson, Mississippi, and thereafter devoted her life to missionary work (Michigan *Alumnus,* 1899). She was dropped from GSA rolls in 1894 for nonpayment of dues.

It must be noted that both Miss Holmes and Miss Florence Bascom, who appears below, were elected to GSA Fellowship immediately after receiving their doctorates. This would not have been possible a few years later, regardless of their sex. As discussed above, GSA began by welcoming all "working geologists" to its ranks but gradually demanded more and more experience and productivity

from its Fellowship candidates. Even R.A.F. Penrose, Jr., might well have been excluded from Fellowship when he was nominated in 1889, only three years out of Harvard!

Florence Bascom, believed by most to have been the first woman in the United States to be recognized as a professional geologist, received her Ph.D. from Johns Hopkins in 1893 and was elected to GSA Fellowship in 1894, five years after Miss Holmes. She was acclaimed throughout a long and active life as "the dean of women geologists." This deserved reputation came not only for her own field and laboratory work at Bryn Mawr and the U.S. Geological Survey on the metamorphic and igneous petrology of the Appalachian region but also because of her ability to instill other women with her own enthusiasm and knowledge. Several of the most outstanding women geologists of the first half of the present century, Ida Ogilvie, Eleanor Bliss (Knopf), and Anna Jonas (Stose) were among Miss Bascom's disciples. Her influence on geology was great, not only through her own research but also in producing inspired women students who could compete with and surpass many men in what was long considered to be a man's world.

Within the Society, Miss Bascom was recognized by election to the Council for the 1925–1927 term and to the Second Vice-Presidency in 1930 (the same year that R.A.F. Penrose, Jr., served as President). These recognitions were small, however, as compared to the honors that came to her from many other sources and many countries. Among the many tributes to her lifetime contributions to geology, those of Knopf (1946) and Ogilvie (1945) are outstanding.

During the Society's first 58 years, through 1946, a total of 28 women scientists were elected to Fellowship. These were:

1889 Mary E. Holmes
1894 Florence Bascom
1907 Ida Helen Ogilvie
1913 Mignon Talbot
1919 Eleanora F. Bliss Knopf
        Marjorie O'Connell Shearon
1920 Julia Anna Gardner
        Carlotta Joaquin Maury
1921 Winifred Goldring
1922 Anna Isabel Jonas Stose
1929 Alva C. Ellisor
1931 Margaret Fuller Boos
1932 Fanny Carter Edson
1934 Grace Anne Stewart
1935 Katherine E. H. Palmer
1936 Alice Evelyn Wilson
1938 Katherine Fowler-Billings
        Madeleine A. Fritz
1939 Helen J. Plummer
1941 Christina Lochman-Balk
1943 Jewell J. Glass
1945 Hildegarde Howard
1946 Esther J. Aberdeen
        Alice S. Allen
        Esther E. R. Applin
        Anna Heitanen-Makela
        Helen N. Loeblich
        Helen M. M. Martin

Two women, Emilie Jäeger (1979) and Dorothy Hill (1980), have been elected as Honorary Fellows. Fewer than 100 other women have become Fellows since 1946, but hundreds have been elected to membership since 1947 when the Society first opened its doors to Members as well as Fellows. The names or exact numbers of Society women, either Fellows or Members, must remain unrecorded here. In fear of unjust charges of discrimination, the Council in 1972 ordered that membership records must contain no listings by sex or race. One could, of course, arrive at lists by scanning the Membership Directories and the annual lists of new Members and Fellows, but this would defeat the wishes of the Council and the results would be imperfect because so many given names are used for both males and females.

Hasty sampling of identifiable women's names in the 1977 *Yearbook* indicates a female membership of between 600 and 1,000 in a total of about 12,000 members. Women thus seem to join GSA in roughly the same proportion as

they bear to men in the geologic population as a whole. The numbers elected to Fellowship, however, are disproportionately small as compared to the female population within the Society. There has been no discrimination against women in the awarding of research grants; the many women applicants for such grants through the years have fared at least as well as men.

Several of the women elected to Fellowship have never contributed largely to original research, nor have they had extensive bibliographies of published papers. They have thus lacked the credentials that come to most minds as absolute requirements for the honor of Fellowship. This is not said in disparagement of the women in question. Each of them has contributed substantially to the welfare of the Society, to geology, or to both. Their elections were based on a single one of the criteria adopted by the Council for eligibility to Fellowship (*Council Rules, Policies, and Procedures*, 1974, p. 16). This says in part: "In exceptional circumstances, candidates may qualify for Fellowship through outstanding service to the geologic sciences that does not fall strictly under the preceding categories, such as extended service in bibliographic or editorial work in the field of the geologic sciences, involving abundant subject-matter knowledge and constant awareness of newly published research results." There can be little doubt that this specific phraseology was originally chosen to fit the accomplishments of specific candidates.

Women as Officers

Comparatively few women have been elected to serve as officers of the Society. Those few were as follows:

1925–1927 Florence Bascom, Councilor
1930   Florence Bascom, 2d Vice–President
1950   Winifred Goldring, 3d Vice–President*
1953   Julia A. Gardner, 3d Vice-President*
1962–1963 Agnes Creagh, Executive Director and Councilor
1965–1967 Agnes Creagh, Secretary to the Council
1974–1976 Helen L. Cannon, Councilor
1975–1977 Joan R. Clark, Councilor
1979–1981 Helen Tappan Loeblich, Councilor
*Ex officio, as Presidents, Paleontological Society

In addition to these elected officers, a few women have been appointed to serve on various committees, particularly in recent years. All who have served the Society in any capacity have done so conscientiously and well. The reasons for the weak representation of women in the management of the Society are not hard to find. First, until recently comparatively few women chose geology for their life work. Second, for numerous reasons that need not be discussed here, even fewer women geologists in proportion to their population in the profession have achieved the scientific stature that could bring them to the attention of nominating committees. Third, and perhaps most important, a significant number of those women who have been approached for service as officers or on committees have found it necessary to refuse. This last reason, unfortunate as it is, is based on the fact that present-day outstanding women scientists, like outstanding members of racial minorities, are so besieged by calls for their services that they have to refuse many requests.

Subdivisions of the Society

Despite its life-long commitment to represent all facets of the earth sciences, GSA has also encouraged and nurtured the growth of specialties within the broad field. Long ago it gave birth to two daughter societies, the Paleontological Society (1910) and the American Mineralogical Society (1920). GSA supported both of these and their publications for many years and has always shared its Annual Meetings with them. Of the many other specialistic groups that have grown since 1888, all have been welcomed to the science and some meet regularly with GSA as Associated Societies (see Chapter 16).

Divisions

In addition to encouraging independent societies, GSA has encouraged the growth of specialistic disciplines by organizing divisions within itself. Divisions are voluntary groups of individual GSA members who are interested in particular disciplines. No criteria for establishment of divisions, either as to size or purpose, have ever been considered or adopted by the Council. Informal groups of interested specialists have customarily petitioned the Council for permission to form a division. A set of bylaws, compatible with the Society's Constitution and Bylaws and commonly drawn up with advice from Society headquarters, is a necessary preliminary to Council approval. Affiliation by the GSA membership with one or more divisions is available for the asking but currently involves a token fee, the proceeds of which are used to produce newsletters (published by the Society for the division members) and for other small expenses.

All divisions hold annual and technical meetings in conjunction with the GSA annual meeting, and some organize separate field trips at that time. Administration is similar to that of the geographic sections with small groups of officers and management boards elected by the affiliates under the Bylaws ratified by the Council. Guidance by the parent Society is minimal, but the divisions are perforce a little less independent than are the sections. Most divisions, for example, have established award funds of some kind with money contributed by affiliates or other donors. These are administered by the individual divisions, but they must be coordinated with the other award programs of the So-

ciety. Through representation on the Joint Technical Program Committee, the programs of each division, field trips, and technical sessions are coordinated with the overall annual meeting program. The Engineering Geology Division long ago took the responsibility for two series of book publications—Reviews in Engineering Geology and Engineering Geology Case Histories. Other divisions will probably follow suit in time. The Engineering Geology Division assembles the raw material for its volumes, but the staff at headquarters takes care of all the details and expense of publishing and marketing the final products.

The Society's divisions are as follows: Engineering Geology, April 1947; Coal Geology, April 1953; Quaternary Geology and Geomorphology, April 1955; Hydrogeology, April 1960; Geophysics, October 1971; History of Geology, November 1976; Archeological Geology, May 1977; and Structural Geology and Tectonics, November 1980.*

Regional Sections

The Society has six geographic or regional sections which together cover the United States and Canada. From time to time there have been discussions about including Mexican geologists in the sectional scheme, either by establishment of a separate section or by extending the boundaries of one of the three existing sections that adjoin Mexico. The discussions have not led to positive action largely because GSA members who reside in Mexico are few and diversified in their interests. Mexican geologists are of course welcome at any meeting of established sections, as are all others.

The sections operate under almost uniform sets of Bylaws that are ratified by the GSA Council, and for tax reasons their finances are handled and accounted for at GSA headquarters. Otherwise, they are autonomous (sometimes fiercely so). In 1976 the Council authorized the sections to incorporate as separate entities if they so desired; this was designed primarily to eliminate some of the tax and accounting problems that were inherent in central management by headquarters. By 1980 none of the sections had taken advantage of the proffered freedom.

The sections elect their own officers and program committees and finance themselves through registration fees at their annual meetings. Section chairmen are changed annually, and by custom most are chosen from among the geologists at the university or city that will host the next meeting. Management boards also serve one-year terms, but secretary-treasurers commonly serve several years. Management and activities are far more informal than those of the parent Society. Section memberships generally

---

*Editor's note: The Planetary Geology Division was formed in May 1981.

consist of all GSA members who reside within the section boundaries, but all sections have provisions for those geologists who reside in one section but have geologic interests in another. Members are informed orally of activities at brief business meetings during annual meetings and by separate editions of *Abstracts with Programs* for each of their meetings. The GSA Council receives annual written reports from the section secretaries.

The major activity of each section is to hold an annual meeting, with technical papers, field trips, and the usual social events. Most meetings are hosted by geology departments at various universities. Earlier activities took place on the host campuses, but attendance at some has now grown so large that they must use convention hotels. Programs and abstracts of the papers to be given at any sectional meeting are published in GSA's *Abstracts with Programs* series (see Chapter 10); some sections also produce guidebooks for their field trips.

Development of the geographic sections has broadened GSA's scope and scientific impact immeasurably. Sections provide opportunities for geologists with common geologic interests to get acquainted and to report, hear, and discuss problems that are of great local interest but not broad enough to fit properly into the annual meetings of the Society itself. Above all, they permit innumerable students and working geologists to communicate with fellow geologists even if they are unable to get to the annual Society conclaves. Several sections make special efforts to involve and encourage the next generation of geologists by giving prizes for best student papers or by other means.

In 1976 the Council gave $1,000 to each of the six sections for use in encouraging students to go on in geology and to join the Society as Student Associates. Sections had complete freedom as to how to use the money. Reactions were mixed. Several sections doled out all the money and reported enthusiastic response from the students, others were only able to use part of the stipend because of shortness of planning time, and still others returned the checks unused. Though the idea of using Society funds to draw the sections and their students closer together was admirable, the execution was perhaps too precipitate; the project was abandoned after a brief trial.

Inclusion of all of the United States and Canada in the sectional scheme arose through a long evolutionary pro-

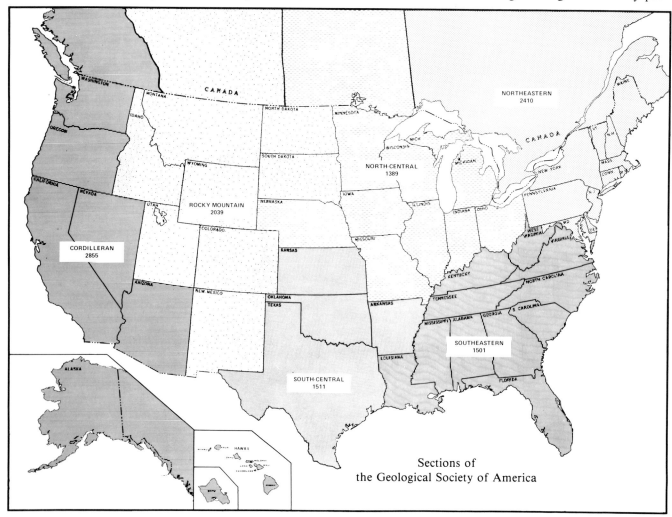

Sections of
the Geological Society of America

cess. Some sections arose spontaneously, others were manufactured. The Cordilleran Section is the oldest by far, and in many respects the strongest and most independent of the sections. Formally authorized on August 26, 1901, when GSA itself was but 13 years old, it originated from a petition in 1899 to the Council, signed by Joseph LeConte, Andrew Lawson, and several other of America's most illustrious geologists who lived and worked on the West Coast. Its first organizational meeting, held at Berkeley, December 29–30, 1899, included presentations of 11 scientific papers. The chief argument for establishment of a section had to do with the difficulty, time, and expense of travel from the West Coast to attend the Society's annual meetings. Rail travel was slow, and the fact that GSA meetings in those days were always held between Christmas and New Year's Day worked additional hardships on both family and professional relationships. The bare statement that the section was established in 1901 as a result of a petition to the Council seems innocuous enough. As can be sensed in the Council Minutes of the period, however, Council's approval of the section averted what might well have come to secession of a large and influential segment of the Fellowship.

A Northeastern Section, centered on New York City, was established with Council approval in 1945. It sank without trace and without explanation within a year or two and has long since been forgotten by most. The Rocky Mountain Section, the first permanent one after the Cordilleran, was approved in April 1947 after a petition from a local group of GSA Fellows. The Southeastern Section, also initiated locally, was established in December of the same year, but it did not get fully organized and hold its first annual meeting until 1952.

The remaining three sections, Northeastern (approved April 1965), North-Central (1966), and South-Central (1966), were initiated not by local groups but by direct action of the Council, which decided that the time had come for expansion of coverage to include all of Canada and the United States. Then Executive Secretary Becker inherited most of the work of organizing local groups and persuading them to prepare the necessary bylaws and petitions and planning for annual meetings.

In recent years, the Society's President or his delegate from the Executive Committee attends each of the section meetings and meets with its officers. As a result of widespread complaints from the sections as to awkward relations between them, headquarters, and the parent Society, the Council began in 1975 to invite each section to send representatives to the Council meetings where they can observe all of the actions and can have an opportunity to air their individual or collective complaints. By and large this plan has had salutory results. These evidences of the parent's interest in the welfare of its children have gone far to mollify feelings of estrangement that have risen from time to time.

The possibility of involving the sections directly in the GSA annual meetings, perhaps by rotating them among the sections' territories and giving sections representation on the Joint Technical Program Committees, has recurred at intervals and has been thoroughly discussed at times. No positive actions have resulted. Not all sections have favored the suggestion, and the very informality and independence that is so treasured by them would present difficulties in recruiting local committees that nowadays must be organized and kept in operation for very long periods before each meeting.

Governance and Leadership

Governance of the Geological Society of America is similar to that of most other societies. It is also similar to most business corporations except that top managements of these usually have seats on the companies' boards, whereas they do not in GSA. Legally, all Society affairs and property are managed by the Council (board of directors) elected by and responsible to the members (stockholders). The Society's officers, also elected by and from the membership and also on the Council, are the President, Vice-President, and Treasurer. Unlike most organizations, GSA no longer has an elected Secretary, but the Bylaws (Art. II, Section 8) provide that the Executive Director, an employee, shall serve as Secretary to the Council.

The Council's chief duties are to set guiding policies and to monitor all Society affairs. In performing these duties it is aided by committees which it appoints as needed. Execution of the Council policies and performance of all the Society's operations are the responsibilities of the Executive Director and the employee staff closely supervised by the Council and Executive Committee (see Chapter 8). The governance pattern just described has evolved slowly and been much changed in detail from time to time, but it is basically the same as that adopted by the Society's founders. That is, GSA has always been managed by a group of officers and a Council elected by and responsible to the membership and assisted by appointed staff, paid or unpaid.

The above sketch of the interrelations between GSA's governing and managing components is in strict accordance with the documents on which the Society is based. That is, the management system as described meets all requirements of both corporate law and of the Society's own Constitution and Bylaws. In actual practice, however, significant departures from strict interpretations of the Constitution have occurred from time to time. From 1923 through 1960, Secretaries Berkey and Aldrich each ran the Society almost single-handed, although the Council was still the legal governing body and was carefully asked to ratify all secretarial actions. During several critical periods of their regimes, moreover, particularly strong Councils and their officers assumed firm control and took the initiative directly. The post-war opening of the Society to a membership class, the major change in investment policies, and the unseating of the Secretary from the Council, all described on other pages, are good examples of this trend.

For the past 10 years or so and increasingly since about 1973, the Executive Committee has gradually tended to take over far more management power than it had in earlier years. This development is described in the section on the Executive Committee.

Legal Basis of the Society

The legal documents under which the Society operates as a not-for-profit organization are the Certificate of Incorporation, the Constitution, and the Bylaws. The Constitution and Bylaws are reproduced here (Appendix B). The Certificate of Incorporation, as amended, is on file at headquarters and with the New York Department of State; it appears in all editions of the booklet *Council Rules, Policies, and Procedures*. Up-to-date versions of the Constitution and Bylaws appear annually in the *Membership Directory*, available for purchase by members and by some libraries.

Certificate of Incorporation

Incorporation had been considered as early as 1913, but the Society did not actually incorporate until 1929 when the Certificate of Incorporation of The Geological Society of America, Inc., together with a new set of Bylaws, was adopted under the membership corporations law of New York (*Bulletin,* v. 41, p. 63). The certificate has been amended in minor particulars several times since 1929. The Society is now also registered as a corporation doing business in Colorado, but it is still under the jurisdiction of the New York Department of State and maintains a statutory office in New York City.

Legal incorporation of the Society was not considered necessary during its earlier years because it was a small organization that depended largely on membership dues for its existence. The move to incorporate in 1929 was originated—and insisted upon—by R.A.F. Penrose, Jr. Evidently even then he was planning to provide for the Society's support and wanted to make it legally possible for it to accept gifts or bequests. Even had there been no Penrose and no Penrose bequest, however, the Society would still have had to incorporate eventually, for modern tax and other laws make incorporation almost mandatory for all not-for-profit organizations, big or little, rich or poor.

Constitution and Bylaws

A provisional Constitution, intended only for organizational purposes, was adopted December 27, 1888 (*Bulletin,* v. 1 p. 7–8; see also Fairchild, 1932, p. 80). A year later it was revised and more detailed and explicit Constitution and Bylaws were adopted (*Bulletin,* v. 1, p. 536, 571–578). This Constitution, reproduced as Appendix A, served the Society well with few significant alterations for nearly three-quarters of a century. The changes up to 1960 are summarized in the *Proceedings for 1960* (1962, p. 161). In a much needed effort to clear up the confusions and ambiguities that by then beset the Bylaws, the 1961 Council appointed an ad hoc committee to recommend needed revisions in the Society's basic documents. It was headed first by A. Rodger Denison, who had long served the Society with distinction, notably as chairman of the Policy and Administration Committee. After Denison's death in an airplane accident, Robert F. Legget took over the chairmanship. Other committee members were J. Edward Hoffmeister and Elburt F. Osborn. By November 1962 the group was able to present a completely revised Constitution and Bylaws to the Council. Among many other significant changes, this Constitution gave Members the right to vote for the first time, gave the Council the right to change Bylaws, and made the Executive Secretary's position appointive and with no seat on the Council instead of elective. Following considerable discussion of Chairman Legget's painstaking explanation of each item, these documents were approved in April 1963 without essential change by the Council and were formally adopted by the Society on November 9, 1963; they did not take effect until 1965. The 1963 Constitution has survived to the present day; it appears as Appendix B. The Bylaws, on the other hand, have been changed many times, mostly in minor ways.

Unlike the Constitution which can be amended only by majority vote of the voting membership, the Bylaws can be amended directly by the Council (subject to subsequent repeal by a majority of the voting membership) or by petition from and vote by the membership. In recent practice, the repeal of Bylaw changes by vote of the membership would be difficult to attain; Bylaw changes voted by the Council are a *fait accompli* long before they first appear in print in the annual *Membership Directory.*

Legal Counsel

Since its incorporation in 1929, the Society has always employed legal counsel. Five firms, represented by half a dozen lawyers, have served at various times. Until recently, counsel received only small honoraria or retainer fees. Successive lawyers have attended Council and annual meetings regularly where they were available to advise on legal technicalities. They have also done research as requested, prepared written opinions, and tended to all the other responsibilities required under corporation law as it applies to a tax-exempt organization. Most of the legal counsels have become deeply interested in the organization and have played active parts in all its deliberations.

In 1929 William C. Armstrong, a New York attorney, was engaged by Secretary Berkey to guide the Society through the incorporation process. Allen B. Memhard was the second attorney. From shortly after Penrose's death in 1931, Memhard guided GSA and played a leading part in settlement of the Penrose estate and receipt of GSA's share. Moreover, his interpretations of New York and federal laws provided guidance and constraints on the Society's investment philosophy and policy throughout his tenure. He resigned in November 1949 after more than 17 years of devoted service. He left somewhat unhappy with the Society because the then Council and the new Treasurer in an obvious wish to alter the existing ultraconservative financial policies had sought a "second legal opinion" on investment and tax problems (Minutes for 1949, p. 197).

Hugh Satterlee, whose firm specialized in corporation tax law, became the third legal counselor in 1949. As he approached retirement in 1960, his place was taken by his partner, Rollin Browne, who served until his death in 1970. Browne's place was soon taken by Morrison Shafroth of Denver, and later by Shafroth's son and partner, Frank. Because the Society is still incorporated in New York, it must retain an office address there. The firm of Hardee,

Barovick, Konecky, and Braun of New York City was selected to provide such an address and to consult with the Shafroths on matters of New York law.

The Society's attorneys' opinions and interpretations of applicable state and federal laws have had profound effects on the Society's investment policies, hence on the health and income yield of its endowment. They have also given aid and comfort to the Council and to headquarters management in helping solve innumerable other problems, both internal and external.

Council Rules, Policies, and Procedures

The 1962 committee that drew up the revised Constitution and Bylaws intentionally made both documents broad and simple. It fully expected, however, that all applicable details of the older documents would be readopted by the Council as foundation for its growing and continuously changing assemblage of rules, policies, and procedures. Prior to 1963 this material, in reality a record of significant Council decisions, was scattered through the Council Minutes, the headquarters files, and in an elusive "black book."

Many references to the "black book" of Council rules appear in the Minutes, especially in the Secretary's reports, throughout the 1940s and 1950s. Entitled "The Geological Society of America, Handbook of By-laws, Rules, and Regulations," it was first issued for in-house use in May 1936. The revision of July 1952 is a loose-leaf booklet of about 75 pages with (predictably) a black paper cover; earlier versions have not been found in headquarters files. The book was obviously intended for use by the staff and to answer procedural questions raised in Council meetings: it was probably never distributed even to Council members.

Just after adoption of the new Constitution in November 1963, codification of all existing Council rules was entrusted by the Council to an ad hoc committee staffed by Edwin B. Eckel, outgoing Councilor; Agnes Creagh, Executive Secretary; and Joan R. Holmes, Administrative Assistant. The committee's product was approved by the Council in May 1964 and published later in the same year as *Council Rules, Policies, and Procedures*. This booklet, as well as revisions of it that were largely written in 1970 and 1974 by Dorothy M. Palmer, Administrative Assistant, was distributed to the entire membership. No revision has been published or widely distributed since, but all rule and policy changes made by subsequent Councils are filed at headquarters. There they are used almost daily by the staff and frequently by officers, Councilors, and committee members.

The Officers

Whether acting as members of the governing Council, as a group or as individuals, the elected officers lead and guide the Society in all of its activities. They also tend to be GSA's most visible representatives in its external relations with other societies, governments, and other groups. Currently, there are only three elected officers—President, Vice–President, and Treasurer. For much of its life the Society also elected Secretaries and Editors as officers; for two decades it even had an elected Librarian.

The offices of President and Vice–President are discussed in the following paragraphs. Because their chief responsibilities are as financial managers, presentation of the Treasurers and their accomplishments is reserved for Chapter 8 on money management. It must be said here, however, that the Treasurers have uniformly contributed much more to the Society than in money matters only. Like Presidents and Vice–Presidents they are elected to one-year terms, but unlike the other officers and Councilors, they can be and usually are reelected annually for as long as they are willing to serve. Customarily, they also serve on the Executive Committee. Thus the Treasurers are able to provide more continuity and experience to Council deliberations than can be acquired by other Councilors during their three-year terms.

Presidents

Depending somewhat on their innate modesties, each of the Society's Presidents has taken the presidency for what it is—with the possible exception of the Penrose Medal—the highest honor for achievement in geology that GSA can bestow. Each of them, too, has cheerfully performed his primary duty, that of presiding as chairman of the board of directors, better known as the GSA Council. Most have delivered outstanding presidential addresses at the close of their terms. Each has presented for publication a summary of his year's stewardship and the state of the Society (although some have asked that these be ghostwritten by the staff).

Beyond this, the President's degree of involvement in GSA affairs has run the gamut from limelight-basking to providing genuine leadership and initiative. Nearly all Presidents have fit the latter category, though because so much of the Society's growth and development is slowly evolutionary, it is difficult to identify individual Presidents as having contributed most to its health through GSA's history. Unfortunately, the Society's formal records tend to obscure the leading parts played by the Presidents in Council meetings. This is because the President customarily chairs the meetings, hence seldom has an opportunity to cast a vote himself. Presidential leadership is behind the scenes.

The President's year is a busy one. Probably no President has actually devoted more than a full quarter of all his working days to presidential duties during his year in office, and there are marked differences between Presidents as to how much they get done themselves and how much they can be depended on by the staff—for advice, decisions, and actual product. The differences seem to lie in the degree of priority each President gives to GSA.

As titular leader of the Society, the President not only presides over Council meetings but must represent GSA with many other groups, such as the International Geological Congresses and the American Geological Institute. He attends as many of the section meetings as possible and is regularly invited to represent his learned Society at inaugurations of college and university presidents (though this last duty and honor is commonly delegated to some Society member who lives near the school involved). As chairman of the Executive Committee the President is in frequent touch with its members by letter, telephone, or personal meetings throughout "his" year. In addition to all these activities, the President is in almost constant contact with Society headquarters, especially with the Secretary/Executive Director. He receives copies of all significant policy-type correspondence, he gives guidance and advice on all manner of problems, and he passes judgment and renders decisions frequently.

Roll of Presidents

The list of Society Presidents contains many of the more illustrious names in geology (see accompanying table and App. D for photographs). Many others of equal or greater stature were never honored with the office. Not a few of the older founders, who had made their marks in the science before GSA was born, died before they could be elected to GSA's highest office. More recently, the science has grown so rapidly in numbers and in scope that at least one new President every month would have been needed to come close to honoring all those who have achieved great heights.

Even so, the list of Presidents is representative of the leaders in North American geology over the past century. The men elected have come from all walks of geologic life and all parts of Canada and the United States. They include explorers, teachers, state and federal officials, consultants, company and independent mineral producers, and investors. Moreover, they represent, as a group, both pure and applied geology in most of the specialties that now characterize the earth sciences.

Vice-Presidents

From the beginning, the Constitution called for election of a President and of two Vice-Presidents; this custom remained unbroken until 1955. The duties were those normally assigned to Vice-President—primarily to serve as backups for the Presidents. A few Presidents were elected direct, but until about 1900, Presidents customarily pro-

gressed upward, serving first as Second Vice-President, then as First Vice-President, and finally as President. This custom gave each of them three full years of experience in helping manage Society affairs. Later, the Second Vice-Presidents were dropped from the Presidential succession (although they were still elected to one-year terms), but the same general objective was maintained with a new custom whereby the outgoing President was elected for a one-year term as Past President and Councilor. This method persisted until the 1963 revision of the Constitution and Bylaws when the title of Past President was removed from the list of officers. The outgoing President, however, is still habitually elected to the Council and is commonly appointed to the Executive Committee. His automatic election to the Council is assured by the clever device of nominating and electing the presidential candidate for one year as President and two years as Councilor (Bylaws, Article II, paragraph 4).

Shortly after the Paleontological Society and later the Mineralogical Society of America were organized and affiliated with GSA in 1910 and 1920, respectively, their Presidents were added to the parent Society's list of officers and Councilors as Third and Fourth Vice-Presidents, ex officio. Through the years these additional Vice-Presidents, who were usually the incumbent Presidents of their own societies, sat regularly with the Council. They had all the same rights as other Councilors except for progression toward the presidency and were solemnly elected by the GSA membership even though in ex officio status. So far as can be seen in the Minutes, few of these "outside" officers took very active parts in Council deliberations.

Finally, a Special Committee on Nominations and Elections recommended to the Council in 1953 that the Society drop the ex officio officers and elect only one Vice-President in the future. The recommendation and the proposed Bylaw change that accompanied it were promptly adopted by the Council, almost without recorded debate. The change did not take place until 1955 because a full set of Vice-Presidents had already been elected for 1954 at the November 1953 corporate meeting. Reasons for the sudden change were never fully defined, but they had to do with fears that with proliferation of specialistic societies and affiliate groups, the GSA Council might sooner or later come to be influenced or even dominated by Vice-Presidential Councilors who were not even members of GSA. The Bylaw change that discontinued the three Vice-Presidents also provided for continuation of a full 17-member Council by adding to the number of elected Councilors.

In theory, the custom of elevating a President to office by stages or of retaining an outgoing President for an additional year is good for the Society. In either case it commits a designee to a three-year term of service, the same as other Councilors, and assures the Council and succeeding officers of the President's accumulated experience and wisdom. In

PAST PRESIDENTS

1889	James Hall*†	1936	W. C. Mendenhall†
1890	James D. Dana*†[1]	1937	Charles Palache†
1891	Alexander Winchell*†[1]	1938	Arthur L. Day†
1892	G. K. Gilbert*†	1939	T. Wayland Vaughan†
1893	J. William Dawson†	1940	Eliot Blackwelder†
1894	T. C. Chamberlin*†	1941	Charles P. Berkey†
1895	N. S. Shaler*†	1942	Douglas Johnson†
1896	Joseph Le Conte*†	1943	E. L. Bruce†
1897	Edward Orton*†	1944	Adolph Knopf†
1898	J. J. Stevenson*†	1945	Edward W. Berry†[2]
1899	B. K. Emerson*†	1946	Norman L. Bowen†
1900	G. M. Dawson†	1947	A. I. Levorsen†
1901	C. D. Walcott*†	1948	James Gilluly†
1902	N. H. Winchell*†	1949	Chester R. Longwell†
1903	S. F. Emmons*†	1950	William W. Rubey†
1904	J. C. Branner*†	1951	Chester Stock†[3]
1905	Raphael Pumpelly*†	1952	Thomas S. Lovering
1906	I. C. Russell*†	1953	Wendell P. Woodring
1907	C. R. Van Hise*†	1954	Ernst Cloos†
1908	Samuel Calvin*†	1955	Walter H. Bucher†
1909	G. K. Gilbert*†	1956	George S. Hume†
1910	Arnold Hague†	1957	Richard J. Russell†
1911	William M. Davis*†	1958	Raymond C. Moore†
1912	H. L. Fairchild*†	1959	Marland P. Billings
1913	Eugene A. Smith*†	1960	Hollis D. Hedberg
1914	George F. Becker*†	1961	Thomas B. Nolan
1915	Arthur P. Coleman†	1962	M. King Hubbert
1916	John M. Clarke†	1963	Harry H. Hess†
1917	Frank D. Adams†	1964	Francis Birch
1918	Whitman Cross†	1965	Wilmot H. Bradley†
1919	J. C. Merriam†	1966	Robert F. Legget
1920	I. C. White*†	1967	Konrad B. Krauskopf
1921	James F. Kemp*†	1968	Ian Campbell†
1922	Charles Schuchert†	1969	Morgan J. Davis†
1923	David White†	1970	John Rodgers
1924	Waldemar Lindgren†	1971	Richard H. Jahns
1925	William B. Scott†	1972	Luna B. Leopold
1926	Andrew C. Lawson†	1973	John C. Maxwell
1927	Arthur Keith†	1974	Clarence R. Allen
1928	Bailey Willis†	1975	Julian R. Goldsmith
1929	Heinrich Ries†	1976	Robert E. Folinsbee
1930	R.A.F. Penrose, Jr.†	1977	Charles L. Drake
1931	Alfred C. Lane†	1978	Peter T. Flawn
1932	Reginald A. Daly†	1979	Leon T. Silver
1933	C. K. Leith†	1980	Laurence L. Sloss
1934	W. H. Collins†	1981	Howard R. Gould
1935	Nevin M. Fenneman†		

[1]Winchell served as Acting President for 1890 as Dana and John S. Newberry, First Vice-President, were both ill. Winchell himself died February 9, 1891, less than two months after his election to the presidency. Gilbert, then First Vice-President, took the chair for a year and was later elected President for 1892.

[2]Berry died September 19, 1945. Bowen was Acting President to the end of the year and was then elected President for 1946.

[3]Stock died December 7, 1950, a few days after his election. Thus Lovering served as Acting President for all of Stock's term and was then elected President for 1952 in his own right.

*Original Fellow
†Deceased Fellow

practice, the interest and activity of some Presidents drop off markedly when they step down; even so, the scheme seems to be the best and most workable that has been devised to get the most benefits from the Society's leaders.

Secretary (Executive Director)

For three-quarters of a century the secretary was an elected officer and Councilor. The revised Constitution and Bylaws, adopted in 1963 and effective from the beginning of 1965, dropped the Secretary from elected office and as a Council member—and gave him a different title. Whether paid or unpaid, and whether called Secretary, Executive Secretary, or as now, Executive Director, the Secretary has unquestionably always had more influence on the health, welfare, and directions of the Society than any other officer. This is so partly because most Secretaries, like the Treasurers, stay on the job for some years. In greatest part, however, it is because the Secretary has almost total responsibility for the day-by-day running of the Society. He follows the policies as laid down by the Council, and although the President, Executive Committee, or others oversee his work, he is the one who gets the Society's work done.

Because the secretaryship is virtually synonymous with management of the Society, further discussion of the job and introduction to the individuals who have held it are reserved for Chapter 7 on headquarters management.

Other Officers

In 1897 the Editor was added as an elected officer and Councilor. This practice continued through 1931 when Secretary Berkey, later aided by Assistant Secretary Henry Aldrich, took over the editorship from Joseph Stanley-Brown. The elective office and official title of Editor was abandoned, but the duties and responsibilities were continued as part of the Secretaries' jobs. Not until 1960 did the Society reestablish the separate job of Editor, and from then on he has been an employee rather than an officer.

As noted in Chapter 6, the Society's list of elected officers included a Librarian from 1897 through 1918 (Fairchild, 1932, p. 197). Prior to 1897 the Secretary was Acting Librarian as part of his duties; the office of Librarian was abandoned in 1918 long after the Society had disposed of its library. Council Minutes of the period indicate that because the Librarian was an elected officer, he had a regular seat on the Council.

Council

The Council, supreme governing body of the Society, has changed in complexion and in numbers many times but not very much in duties or responsibilities. The Provisional Constitution, adopted in 1888, called for an Executive Council, composed of a President, two Vice-Presidents, a Secretary, a Treasurer, and 3 additional Fellows, all elected by the Fellowship. The term "Executive Council" was soon shortened to "Council," and 3 more Fellows were added. This brought the roster to 11 councilors—5 officers and 6 at-large members.

The size of the Council has varied through the years, depending partly on whim and partly on the number of elected officers. Most recently the Bylaws set Council limits of 10 to 24 members; currently the Council is made up of 3 officers and 13 others (including the Past President) or a total of 16.

For many years, New York law required at least one Director (Councilor) to be a resident of New York State. This restriction troubled nominations committees at times, especially so when the rush to suburbia led many eligible geologists to move from New York City and its suburbs to adjoining states. The requirement is no longer in effect.

Regular councilors serve three-year terms on a staggered basis, four being elected each year. Officers are elected for only one-year terms, but Vice-Presidents ordinarily move up to President and automatically to Past President and Councilor. Treasurers are normlly reelected annually for as long as they are willing to serve. As noted above, the Society no longer has an elected Secretary, but the Executive Director fills all legal requirements as Secretary to the Council. He sits with the Council and commonly with the Executive Committee but has no vote on either.

In the early days, the Council could dispose of its business for the entire year in less than half a day. Meetings were held during the annual meeting of the Society. Since about 1920, Councils have generally met twice a year with spring meetings at headquarters and the fall ones during or just before the annual meeting. Time devoted to Council meetings has increased with the growth in Society size and complexity of its problems. For many years the Council has required at least two full days to complete its work. Even this schedule would be greatly extended were Council not provided with a large "notebook" for advance study. Quite different from the "black book" of Council rules mentioned previously, the "notebook" contains virtually all committee reports, recommendations for action, and other items that are to be discussed. The Council proceeds on the assumption that all Councilors have done their advance study so that material in the notebook need not be repeated orally. In practice, not all officers or Councilors do their homework properly, but peer pressure generally prevents them from admitting their shortcomings in open meeting. Aided by the notebook device and by the Executive Committee's prior action on many items, the Council is able to dispose of an enormous amount of business over a couple of days.

During the Society's first 30 years or so, Council meetings were relatively informal; they were also likely to be catch-as-catch-can affairs with multiple brief sessions sandwiched in between the Society's technical meetings, just before or after meals, or even in late evening hours. With time, the meetings became more formal and more rigidly scheduled. This trend first became apparent around 1923 when C. P. Berkey became Secretary and began to apply his executive abilities.

The President (or the Vice-President in his absence) presides over all Council meetings. Business is conducted according to *Robert's Rules of Order*, and an agenda prepared and distributed in advance is followed, though with modifications to fit the needs of the moment. Most decisions are made by voice votes, but nominations, appointments to committees, and election of medalists and Honorary Fellows are commonly made by secret written ballots. Most committee reports are presented in person by their chairmen, and key staff members are usually asked to sit in during Council study of their areas of competence and to answer questions. In recent years representatives of each section are also invited to attend as observers but very seldom as participants. Except for closed sessions, all Council discussions and decisions are recorded in shorthand and on tape by staff members, who produce and distribute the minutes soon after adjournment. After editing and approval, these become the official records of Society actions and policies. The degree of detail recorded through the years has ranged between wide limits, depending on personalities, secretarial skills, and advances in recording equipment. By and large, however, the written record of Council actions is comprehensive and of excellent quality. Although they may take time for an occasional joke and a round of laughter, Councils have been uniformly conscientious and business-like throughout the Society's history, and their members have given much of themselves in providing sound guidance to the Society.

At some point during most Council meetings, "closed" or "executive" sessions are held. Staff members and visitors are excused, leaving the Council or Executive Committee free to discuss in private staff performance, salary scales, or other sensitive subjects. Decisions that result from closed sessions and that require staff action are usually transmitted orally by the President to the Secretary/Executive Director, but written records tend to be sparse.

For many years, incoming Councils have held informal meetings immediately after the outgoing Councils have concluded their terms and adjourned. The main purpose of the informal meetings has been to organize the incoming Council and to appoint committees for the coming year. The timing is convenient because two-thirds of the old Council members will immediately become part of the new Council, and all have the Society's problems fresh in mind. After the annual election of officers and Councilors, the new Council can hold its first formal meeting, ratify the actions of the informal meeting, and go on to new business.

In addition to closed sessions and the customary informal meetings—"informal" only because the new Council cannot take any legal action until it has been officially elected at the annual corporate meeting—Councils from time to time have held general, off-the-record discussions. These may cover nebulous or incipient ideas toward betterment of the Society's activities or solution of a specific problem, or they may examine some facet of new geologic knowledge. At times, these gatherings have been held the evening before the Council meeting; at other times they have occupied the last hour or two before adjournment.

"No Council Can Commit—"

Each new Council inherits a saying, which has become almost a cliché, to the effect that "no Council can take any action that will commit a future Council." This belief is patently based on false premises. Any Council may, of course, rescind any action of a previous Council, just as many Councils may—and frequently do—reconsider and rescind their own actions. But many, if not most, Council actions are really long-term commitments whose effects must carry on from one Council to the next to have any meaning. Such actions as nominations of officers, elections of medalists, changes in the Bylaws, investment or publication policies, commitments for future annual meetings, establishment of annuity and other fringe-benefit plans, and purchases of real estate, for instance, are all ones that must have some promise of stability over a period of years. To the credit of successive Councils, few decisions of comparable importance to these examples have ever been made—or rescinded—lightly. But veteran Council-watchers through the years have seen many lesser decisions made hastily and capriciously in the evident belief that more thought would be wasted because "No Council Can Commit. . ."

Occasionally, a decision is made on the spur of the moment and with no warning, almost like the clear-air turbulence that is so well-known to airplane pilots. Official Minutes seldom record the backgrounds of these sudden decisions, but those who have observed them can usually ascribe them to a single speaker, one who speaks impulsively but so forcefully (and commonly loudly) that he carries his issue to a majority vote in a matter of minutes. There is one such in almost every Council.

One minor but illustrative action took place at the fall Council meeting in 1972. It arose during discussion of such an innocuous item as the routine appointment of two representatives to the American Association for the Advancement of Science. Before it knew it, the Council had agreed not only to refuse to appoint representatives but also to withdraw GSA from association with the AAAS! The arguments for this action, some perhaps real but most of them specious, were that AAAS had tended to ignore the earth sciences in its meetings and in its journal *Science*. (Ironically, the arguments used were almost identical with those that had led to formation of the Society in 1888.) In its haste to disassociate, the Council forgot that it was disowning its own mother, for GSA sprang from the AAAS in 1888. Embarrassing as it was at the time, the dissociation lasted only until the next meeting in 1973 when the Council exercised its time-honored prerogative of changing its mind and asked to be returned to the AAAS fold; AAAS graciously accepted its repentant offspring.

Executive Committee

Until the Society inherited its share of the Penrose fortune, there was no need for an Executive Committee. The Council itself was small and its members could easily act as a unit either at in-person meetings or by mail ballot. The few problems that rose between scheduled meetings were commonly solved by the Secretary, with or without consultation with the President or other officers. Of these, the Editor and the Treasurer were the ones turned to most frequently.

The revised Constitution adopted in 1932 provided for an Executive Committee, authorized to act for the Council between regular meetings. Until 1945, however, the Executive Committee was permitted to work only on specific problems assigned to it by the Council, not to take final actions on matters that were too urgent to wait for Council decisions.

From 1932 through 1941, the first 10 years under the Penrose fortune, the Executive Committee was composed of the President, the Secretary, and the chairman of the Finance Committee. (This custom added immeasurably to the Secretary's influence on the Society because he was also a voting member of the Council.) When W. O. Hotchkiss took over the Finance Committee in 1942 to replace Joseph Stanley-Brown, deceased, Hotchkiss was not appointed to the Executive Committee. Instead, the President, Secretary and Vice-President were asked to serve. This lasted for only a short time, however, and by 1944 Hotchkiss, then Treasurer and chairman of the Finance Committee, replaced the Vice-President. He retained all three positions until his retirement in 1948, but when the new Treasurer, J. Edward Hoffmeister, took office, the Council decided to split the responsibilities. A Vice-President, Wendell P. Woodring, became the third member of the Executive Committee, and Hoffmeister became an ex officio member of the Finance Committee rather than its chairman; Cyril W. Knight was chosen for this post. The size and complexion of the Executive Committee has varied somewhat since 1948 depending partly on changes in the Bylaws and partly on the wishes of incoming Presidents.

According to today's Bylaws, Article 2, paragraph 8 (see Appendix B), the Executive Committee is appointed by the council and is empowered to act for it between regular Council meetings. All its actions are subject to Council ratification. The Executive Committee is to consist of three to five Councilors, including the President (chairman) and Vice-President. Normally, the Past President and Treasurer are also included. The slate of committee members is prepared in normal fashion by the Committee on Committees for election by the Council, but in practice both these groups defer to the incoming President's wishes for the makeup of the Executive Committees. In 1977 a major but unheralded addition was made. The Committee on the Budget, a separate entity, was retained, but the Executive Committee in effect took over all budgeting responsibilities by adding the chairman, Committee on the Budget, as an ex officio member of its own group.

Unquestionably, the Executive Committee has always performed its primary duty—that of acting for the Council between regular meetings—in exemplary fashion and in the best interests of the Society. It meets just before each Council meeting and several other times during the year. Many important decisions have been made at such meetings on matters that could not wait for Council action yet could not justify the calling of special Council meetings.

During the past decade the Executive Committee has gradually assumed far more of the Society's management functions than merely solving emergencies between Council meetings. Aside from its assumption of budget preparation, the most striking evidence of this trend, perhaps, is the fact that before each Council meeting the Executive Committee reviews the detailed agenda and the Council notebook. It makes preliminary decisions on a high proportion of the notebook items and presents the decisions as firm recommendations to the Council. At this point any Councilor may call for tabling of specific recommendations, reserving them for Council action after full discussion. This privilege is seldom used, however, with the result that much of the Council's business is disposed of by a single motion to ratify the Executive Committee's recommendations. The system unquestionably saves much time and energy for the Council and the staff, which is its avowed purpose, but it also tends to stifle freedom of action by the Council and fails to provide for adequate Council-wide discussion of many items of business that would perhaps benefit from more points of view than can be supplied by the Executive Committee alone. In addition to its preliminary actions on Council agendas, the growing dominance of the Executive Committee is shown in many other ways.

De facto management by the Executive Committee has met with surprisingly little resistance. A few representatives of sections, divisions, or committees who have seen the Council in action or come closer to Society management than most individual members have occasionally voiced alarm. Within the Council itself complaints have been expressed from time to time to the effect that the Council has gradually become a "rubber stamp" for the wishes of the Executive Committee. In recent years the minutes show that at least two Councilors have formally expressed strong objections to the dilution of the Council's place in Society leadership. Such expressions have elicited some lively discussion but no visible changes in procedure.

There is nothing intrinsically wrong with policy-setting and decision-making being done by a small Executive Committee rather than by a broader-based and somewhat more unwieldy Council that is more representative of the total membership and intended to be the supreme governing body of the Society. All actions of the kinds discussed here are constitutionally and legally correct, since Executive Committee actions become Council actions, and these become Society actions by routine ratification processes. Moreover, each Executive Committee is made up of leaders in the profession, each with sound judgment and each dedicated to the best interests of the Society and of geology. If there is any cause for alarm in the trend toward dominance by the Executive Committee in Society affairs, it lies in the fact that, unrealized by the electorate, the intentions of the Constitution and Bylaws are being gradually eroded. If management by a small Executive Committee is indeed preferable to that of a larger and more representative Council, amendments of the basic documents would seem to be in order. If this is not the case, future Councils must be more aggressive in asserting their constitutional authority. In any event, it seems desirable that the membership be better informed of policy and management decisions and of their sources than they have been in the past.

Committees

Most policy decisions and other actions taken by the Council and by Society officers are based on recommendations by committees. This procedure is similar to that followed by legislative bodies everywhere. It is based on the realization that no group, no matter how constituted, can possibly have the time, skill, or other resources to gather the necessary information for wise decisions on a very wide variety of subjects. This applies to any such body, whether it rules a nation, a corporation, or a society. Interested individuals or groups naturally have rights to appear directly before the Council on urgent or important matters, but these rights are almost never used.

Committees range widely in size form from one to a dozen or more members. Regardless of size, makeup, or tenure, each committee is an important and integral part of Society management. In fact, the Society could hardly function without them, for neither the Council nor the staff could possibly supply the time, knowledge, or judgment that is generously donated by scores of committee workers year after year. Moreover, GSA's financial resources, large as they are, will never be large enough to buy all the advice that comes to it free from its committees.

Names and makeups of all committees of the Council are listed annually in the *Membership Directory*; functions of each can be perceived in a general way from their titles. Duties and procedural rules of all standing committees are described in *Council Rules, Policies, and Procedures.* The work of a few committees, such as those on Investments, Publications, and Joint Technical Program, is described in other chapters; many others, both permanent and transitory, are mentioned in numerous places.

Some committees, particularly the ad hoc ones, are appointed by the Executive Committee or by the President. Most of them, however, are appointed by the Council, based on lists of candidates proposed by the Committee on Committees. This last is one of the few committees that is staffed afresh each year, the purpose being to bring more new blood into Society affairs and to avoid even the appearance of self-perpetuation on the part of the Council.

For some reason, the very term "committee on committees" has appealed to many members' senses of humor for many years. Those who believe it to be funny seem to think that a Committee on Committees somehow organizes and supervises the work of all the other committees. In reality, the Committee on Committees is a group of dedicated geologists carefully selected for wisdom and for collectively wide acquaintance among the outstanding workers in their fields or parts of the continent. After careful study of the requirements for each committee and of all potential candidates, they prepare a multiple slate of nominees for each open position and present the list to the Council for final selections. No better method of ensuring the continued health of the Society and of avoiding an entrenched bureaucracy than the Committee on Committees concept has been devised.

Committee on Policy and Administration

The Policy and Administration (P and A) Committee functioned effectively from 1945 through 1964 when it was deleted from the revised Bylaws that went into effect in January 1965. It was first called "Organization, Policy, and Administration," then "Operations, Policy, and Administration," and finally "Policy and Administration."

Throughout its life the P and A Committee was one of the most productive and valuable of all the Society's structural elements. Composed of outstanding elder statesmen, almost all with keen minds, vigor, and extreme loyalty to the organization, the P and A Committee characteristically investigated any and all sticky problems that required more time, thought, and intuition than could be applied either by the Council or by its regular committees. From time to time the P and A Committee initiated studies of real or incipient problems on its own; more commonly, however, it was assigned specific problems by the Council without restrictions as to its methods or directions. In this function it in effect took over many of the former duties of the Executive Committee, which prior to 1945 was empowered only to handle problems or take actions specifically assigned to it by the Council.

Reasons for abolishment of the committee were discussed by the Council but do not appear in the records. Apparently the Committee on Revision of the Constitution and Bylaws perceived that the P and A Committee had gradually abrogated some of the Council's authority. In order to restore that authority to its rightful place, it seemed desirable to abolish the P and A Committee rather than attempt to limit its powers. Since the demise of the P and A Committee, problems similar to those it had studied now go to the Executive Committee, to ad hoc committees, or more often than not to the Executive Director and his staff.

7

Headquarters Management

Working under guidelines and policies set by the Council, the Society's day-by-day operations are performed by the Secretary/Executive Director and a paid staff at GSA headquarters in Boulder, Colorado. The same pattern of management from a general headquarters, supervised by the Society's Secretary, has existed since 1888 but in a variety of locales.

Supporting activities are scattered over the entire country. These include the *Treatise on Invertebrate Paleontology* staff at Lawrence, Kansas, paid by the University of Kansas but partly supported financially by GSA. Through 1980, the *Bulletin, Geology, Abstracts with Programs,* and many other items were printed under a long-standing semiformal agreement by the Lane Press, Burlington, Vermont. This firm also stored and distributed all Society publications. A change in these arrangements was under consideration early in 1981. As they have been for many years, most large maps and charts are printed by Williams & Heintz in Washington, D.C. Books and some other publications are contracted to half a dozen printing houses on the basis of competitive bids on each individual item, but a significant part of the miscellaneous printing is done at Boulder headquarters. The Society's two legal counsels are based in New York and Denver, auditing is done by a firm in Denver, and investment advisors and financial custodians are in New York. Computer services are based in Boulder.

In the following pages, the Secretaries on whom the Society has depended through the years are introduced individually. Then their responsibilities and positions in the hierarchy are discussed. Finally, the character and duties of the staff that keeps the Society moving are briefly outlined.

Secretaries/Executive Directors

Ten Secretaries/Executive Directors (plus two Secretaries to the Council) have served the Society during the years 1888–1979 (see accompanying table). Their terms of office have ranged widely, as have their titles. The various titles applied through the years to the Society's Secretaries reflect changes in semantics more than they do changes in character and number of responsibilities. Adoption of the title "Executive Director" was assertedly made because the old wording ("Secretary" or "Executive Secretary") had become archaic in the business world. Although never used by GSA, the title of Chief Executive Officer comes closer to characterizing the top employee of the present-day Society than do any of those that have actually been applied.

John J. Stevenson, the Society's first Secretary, served for only two years, resigning at the end of 1890 and relinquishing the reins to Herman L. Fairchild, but Stevenson played a far larger part in molding the Society than is indicated by his brief tenure. A professor at the University of the city of New York, he maintained GSA headquarters at his office on Washington Square, New York City. A founder and one of the Original Fellows, he became the Society's first Honorary Life Fellow in 1891 and served as GSA President for 1898.

Much of Stevenson's best work for the Society was done during the mid-1880s before there even was a GSA. He was one of the more ardent organizers of the new Society and worked closely with the Winchells, Hitchcock, and others in promoting the cause and in managing the breakaway from the AAAS. As Secretary of the Organizing Committee he wrote most of the material that went to potential members, kept records, and doubtless had a strong hand in drafting the first Constitution and Bylaws. The Minutes and other records of his two years as Secretary are brief and unassuming, but he obviously played a major part in setting the new Society on a sound course. (For Stevenson's memorial, see the *Bulletin,* v. 36, p. 100.)

Herman LeRoy Fairchild was Secretary for 16 years, 1891 through 1906, and President in 1912. He was one of the founding fathers. He taught geology at the University of Rochester, where Society headquarters were located during his tenure. Toward the end of his career he was asked by the Council to prepare a history of GSA. He accepted the assignment with enthusiasm and produced an excellent story of the first 40 years, based almost entirely on his own observations and personal friendships with most of the principals (Fairchild, 1932). By coincidence, his history is of the pre-Penrose era for Penrose's death and his magnificent bequest occurred just as Fairchild's book was going through the press. I have drawn heavily on his history here, but for details and for the flavor of the times, one should read Fairchild in the original. For one of several memorials, see Chadwick (1945).

Edmund O. Hovey was elected to replace Herman Fairchild as Secretary in 1907. He served through 1922 and stayed on as Councilor until his death in 1924. An Original Fellow (and son of another Original Fellow, Horace C. Hovey), he received his geologic education at Yale and Heidelberg and was curator and geologist with the American Museum of Natural History for his entire professional life. Society business was done out of his office at the museum, but it occupied comparatively little of his time, since he was often away on explorations in the Arctic and elsewhere.

As Secretary, Hovey kept the Society on an even keel for many years of comparatively routine operations. His reports to the Council and membership are succinct, and he made frequent excuses that he had "not had opportunity to write" to someone. He took some independent actions but sought Council decisions on most. Hovey's administration was marked by the admission of the first affiliates— Paleontological Society and Mineralogical Society of America—as well as of the first Correspondents (Honorary Fellows), and by healthy growth in membership and publications. On his retirement from the secretaryship, Hovey

ROSTER OF SECRETARIES

Name	Incumbency	Official Title
John J. Stevenson	1888–1890	Secretary
Herman L. Fairchild	1891–1906	Secretary
Edmund O. Hovey	1907–1922[1]	Secretary
Charles P. Berkey	1923–1940[2]	Secretary
Henry R. Aldrich[3]	1941–1959	Secretary
Frederick Betz, Jr.	1960–1961	Executive Secretary
Agnes Creagh	1962–1963	Executive Director
Joe Webb Peoples	1962–1964	Secretary to Council[4]
Agnes Creagh	1965–1967	Secretary of Council[4]
Raymond C. Becker	1964–1969	Executive Secretary
Edwin B. Eckel	1970–1974	Executive Secretary
John C. Frye	1974–19__	Executive Director

[1] For most of 1915–1917, Charles P. Berkey was Acting Secretary while Hovey was on an icebound exploratory ship off Greenland.

[2] James F. Kemp was Acting Secretary during most of 1925 while Berkey was in Mongolia.

[3] Aldrich was the first full-time salaried Secretary; he was also Editor-in-Chief from 1934 until his retirement February 28, 1960.

[4] This position, with no management responsibilities, was necessary because these Bylaws required that the Secretary to the Council must be an elected Councilor.

was given a silver loving cup by his friends at the annual meeting in Ann Arbor, and James F. Kemp read a resolution of appreciation that had been prepared by a special committee of James F. Kemp, John M. Clarke, and R.A.F. Penrose, Jr. (*Bulletin,* v. 34, p. 76–80).

Charles Peter Berkey, educated at Minnesota and Columbia, became Secretary in 1922, but he already had some idea of what the job was like for he had acted for Secretary Hovey for nearly three years during 1915–1917. His own secretary at Columbia, Mrs. Miriam F. Howells, had also helped him during that period so that the two were an already experienced team when Berkey took office.

Berkey taught at Columbia from 1903 through 1941 and was head of the department of geology for many years. He was a pioneer in engineering geology and practiced in that field, with worldwide acclaim, from 1906 almost until his death in 1955 at age 88 (Savage and Rhoades, 1950). With all his responsibilities to the university and to his consulting work he obviously had only a part of his time and energy available for the Society. He was intensely loyal to GSA, however, and put far more of himself into its guidance than is usual for part-time workers. One of several memorials is that by Kerr (1957).

Berkey's influence on the Society was very great. He bridged the period between the old and the new, post-Penrose eras. As discussed in Chapter 3, he almost certainly had more to do with the naming of GSA in Penrose's will than he ever admitted or than most people ever realized. The long-continued friendly ties between Columbia and GSA, though decried by the membership at times, gave the Society a home for 40 years, first in the geology department offices and later in a house of its own; had not Professor Berkey been its Secretary, the Society would surely have evolved elsewhere and possibly differently. Above all else, Berkey's chief claim to credit in the annals of GSA was his smooth but firm guidance during the difficult years of adjustment to sudden wealth, with greatly expanded activities and the setting of new directions.

Berkey crowned his many years of unselfish service to the Society with the presidency for 1941. After completion of his term, he was named Honorary Councilor and as such attended Council meetings for several more years, but the custom was abandoned by Council action long before his death (Proceedings for 1942, p. 9).

Henry Ray Aldrich was employed as Assistant Secretary to Dr. Berkey in 1934. In addition to his other duties he shouldered almost all of the publication problems from the beginning of the expanded publications program and continued as Editor-in-Chief until he retired in 1960. He was appointed to the secretaryship at the end of 1940, becoming the first full-time salaried Secretary the Society had ever had. Aldrich's editorial duties were enormously lightened by Helen R. Fairbanks, who had served as Managing Editor under Berkey from 1932 and who continued in that capacity under Aldrich through 1938. To her belongs much of the credit for launching the expanded publications program.

Educated in mining engineering and geology at Massachusetts Institute of Technology, Minnesota, and Wisconsin and with 15 years of experience in field geology and administration with the Wisconsin Geological Survey, Aldrich had a good background for his work with GSA. His engineering training gave him a liking for and an understanding of figures for budgets, publication statistics, and the like that not all Secretaries have had—and no doubt explains his penchant for charts and graphs which he used in all his reports. A kindly but generally discerning characterization of both the man and his administration is that given by Kerr (1959). Henry Aldrich died January 25, 1979.

The Berkey and Aldrich regimes must be considered almost as a unit, for the two men worked together for six years, and Aldrich continued to build on what Berkey had started. Possibly the two events that highlighted Aldrich's term were the enormous increase in membership numbers that began with opening of the Society to Members as well as Fellows and the major revisions in investment philosophy that followed soon after.

Berkey had taken an increasingly aggressive and direct part in managing the Society's affairs, not only in the day-to-day headquarters activities but also in Council and Executive Committee meetings. Aldrich followed in Berkey's footsteps and, if anything, provided even stronger leadership, to the extent that some members grumbled about one-man rule. Council Minutes of those years give abundant internal evidence of the leadership provided by Berkey and his successor. Theirs are the motions or the seconds to motions, theirs the prepared resolutions ready for presentation at the proper moments, theirs the succinct summaries of wandering discussions with adroit switches to more important matters. Theirs also were innumerable firm decisions and actions taken in the name of the Society between the semiannual meetings of the Council to be reported later to the Council only for *ex post facto* ratification. Aldrich admitted privately to a "proprietary" interest in his position indicating to one of the candidates for his replacement that "the job today [as Secretary] is largely what I have made it." Unquestionably, both Berkey and Aldrich ran the Society with iron hands. If they were dictators, however, both were benevolent dictators of the finest kind.

Both men had well-developed egos, it is true, and no doubt enjoyed a sense of power in running the Society. But they were almost utterly selfless, devoting the better parts of their lives to the good of the Society with no thought of personal gain or aggrandizement. Berkey donated his services with the exception of a small honorarium; Aldrich was paid a relatively modest salary with modest increments, but his remuneration was never large in comparison with his responsibilities.

Secretaries/Executive Directors

John J. Stevenson, 1841–1924; Secretary, 1888–1890; President, 1898.

Herman L. Fairchild, 1850–1943; Secretary, 1891–1906; President, 1912.

Edmund O. Hovey, 1862–1924; Secretary, 1907–1922.

Charles P. Berkey, 1867–1955; Acting Editor, 1932–1933; Secretary, 1923–1940; President, 1941.

Henry R. Aldrich, 1891–1979; Assistant Secretary, 1934–1940; Secretary, 1941–1959; Editor-in-Chief, 1934–1959.

Frederick Betz, Jr., 1915–   ; Executive Secretary, 1959–1961.

In several of his annual reports to the Council and the membership, Aldrich seems to be strongly on the defensive and to feel a compulsion to teach the Council and others some of the facts of GSA life that were so well understood by him. No doubt such reports reflect periods when specially strong Councils were trying to change established customs or to invade the Secretary's sacred precincts.

Both Berkey and Aldrich were characterized by some of their friends as quiet and unassuming, preferring to work in the background rather than in the limelight. Not everyone would have agreed with these views but many did, and whatever their methods, both men got an enormous amount of work from their staffs and from themselves. Hard taskmasters they may have been at times, but every action that they took or asked others to take was for the good of the Society as they envisioned it.

Frederick Betz, Jr., educated at Columbia and Princeton, came to the Society in 1960 from the U.S. Geological Survey where he had had many years administrative and technical experience, mainly in military geology. His was a difficult assignment, not only for him but for the staff and the Society. With only a few days of training, he was faced with taking over a job that had been held for many years by one man, aided by a small staff, most with long service and a deep personal loyalty to the Society. The adjustment to a new and younger leader with different ideas as to the principles of business administration somewhat disrupted the morale of the staff and led to rapid turnover. Replacements

Secretaries/Executive Directors

Agnes Creagh, 1914– ; Editor, 1960–
1964; Executive Director, 1962–1963;
Secretary to Council, 1965–1967.

Joe Webb Peoples, 1907– ; Secretary
to Council, 1962–1964.

Raymond C. Becker, 1906– ; Acting
Editor, 1967; Executive Secretary,
1964–1969.

Edwin B. Eckel, 1906– ; Editor,
1968–1971; Executive Secretary, 1970–
1974.

John C. Frye, 1912– ; Executive Director, 1974– .

for departing staff and a genuine need for increases in numbers and pay of staff in a difficult job market posed challenging problems for the new Secretary. Added to these was another and even more pressing problem—one that had been foreseen for years but that had not been faced realistically. This was the need to find a new home for the Society. Before these and other pressures inevitable at the end of one era and the start of another could be fully resolved, Betz abruptly resigned the secretaryship after less than two years of service.

Agnes Creagh came to the Society as an Editorial Assistant in 1936. Fresh from Barnard College and graduate studies in geology at Northwestern, she enjoyed her work and, under various titles, served as Managing Editor of the entire publications program for 25 years. During most of this time Henry Aldrich retained the title of Editor-in-Chief, but it was Miss Creagh who saw to all the details of maintaining an ever-increasing flood of high-quality publications, just as Helen Fairbanks had before her.

Miss Creagh succeeded Aldrich as Editor when he retired, and not long after when Betz resigned the secretaryship, Miss Creagh was asked to take over, first as Acting Secretary and soon after as Executive Director. Even then she was not relieved of her publications duties, although she was assisted by William C. Krumbein, then on the Publications Committee, and another friend, Fred A. Donath of the Columbia University faculty, who aided in review of manuscripts.

Marked by completion of the transition begun by Retz and the load of her publications duties, her term in charge of the Society's operations was difficult for her. Added to all this, she had to arrange for and oversee the purchase and alterations of a new headquarters building and the move into it. She rose to all the challenges and got the Society moving smoothly again. She did not really enjoy her double job, however, and did not feel equal to its demands. It was easy for the Executive Committee to ignore her repeated pleas for relief, for nowhere else could they find the loyalty, determination, and skills that characterized her entire term in office. Finally, and in desperate realization that she was unlikely to be relieved so long as she was willing to bear the burdens, she give firm notice of her resignation. This she did not only for her own health and peace of mind but for the good of the Society, as she saw the situation.

Raymond C. Becker was educated at Clark and Johns Hopkins Universities. His first ties with the society were in 1931–1933 when he was lent by the U.S. Geological Survey to act as Business Manager for the XVI International Geological Congress. As such, he was involved in planning, logistics, field trips, and lodging of foreign delegates. Later, he worked for many years in a variety of scientific and administrative jobs, partly with the Soil Conservation Service and the USGS and partly in the natural gas industry. He had retired early and was an Arkansas farmer when the Society discovered his availability and asked him to take the reins from Agnes Creagh.

Becker's regime was short but eventful. During the first several years he successfully completed the settling in to the newly occupied New York headquarters on East 46th Street while somehow coping with staffing problems that beset all metropolitan office managers in those years. His greatest challenge and accomplishment, however, had to do with the Society's move from New York to Boulder, Colorado. Not only did he have to find and lease temporary new quarters but he also had to superintend the physical move of a business office and its voluminous records while retaining some semblance of normal operations. All this was done during an almost complete change of staff. Only three of the New York staff chose to make the move, and two of these returned to the East within a few months. All others had to be separated on fair and equitable terms, and an entirely new staff had to be found, hired, and trained in Boulder. This responsibility was exacerbated by a need for a staff greatly enlarged over that of New York to cope with an accelerated publication program financed by a generous grant from the National Science Foundation.

Becker's administration was not without rough spots, but he accomplished the Society's main objectives of uprooting headquarters, reestablishing it in a new environment, and setting it on a new course. The strains took their toll of him personally, however, and he resigned the secretaryship in November 1969 for reasons of health.

Edwin B. Eckel was employed as Editor by the Executive Committee February 1, 1968 (see Preface). When Becker resigned in November 1969, Eckel was asked by the Executive Committee to take over as Acting Executive Secretary and to retain the Editor's duties as well. The word "acting" was dropped from his title a few months later, and he was finally relieved of the editorship in mid-1971.

At least in his own mind, Eckel's regime was marked by smoother employee relations and with greatly improved morale and *esprit de corps* than had obtained earlier. The other and probably more lasting accomplishment during his term of office was his leadership in building GSA's new home, from initial site selection through all the stages of a major construction project to furnishing and occupation. He retired in mid-1974 after 6½ years with the Society, a bit older and wearier, but reasonably satisfied with his stewardship.

John C. Frye became Executive Director September 1, 1974, after many years of outstanding experience in administration of two great State Geological Surveys (Kansas and Illinois); active participation in many other societies, committees, and commissions; and in productive geologic research. Like most of his predecessors, he was immediately faced with problems. Frye's particular emergency was financial—a recession had severely affected the Society's investment portfolio and its income, and inflation was raging. Previous management had kept spending within the authorized budget, but a very large realized loss in the investment portfolio had resulted in an overall deficit in Society finances. Economy was the watchword, and cuts in nearly all phases of the Society's operations were essential. Budget-balancing in times of stress—never a popular activity with anyone but those with the ultimate responsibility for the Society's health—dominated Frye's attention for his first several years. Fortunately, the economic problems had abated somewhat, and GSA could go forward on a sound basis well before this history was ready for the press.

Assistant Secretaries

Several attempts have been made through the years to establish a headquarters position of Assistant Secretary. With two notable exceptions, most of these attempts have been short-lived and disappointing. One exception was Mrs. Miriam Howells, who served Berkey and later Aldrich faithfully and well from 1922 until she retired in 1959. Though both Berkey and Aldrich—among others—referred to her as Assistant Secretary (or Chief Assistant) and although she carried most of the detailed work of the Society for all that time, she disavowed the titles (personal letter to Edwin B. Eckel, November 25, 1977). She was, in fact, what the Society now calls "Administrative Assistant."

The other obvious exception was Henry R. Aldrich. Hired (not elected) in 1934 as Assistant Secretary, he served in that capacity through 1940 when he took Berkey's place as Secretary. From the very beginning of his tenure, Aldrich became concerned with the publication program and was soon named Editor-in-Chief; he continued to hold that title, as well as those of Assistant Secretary and, later, Secretary, until he retired early in 1960.

At intervals from 1929 to the early 1970s, several other Assistant Secretaries were added to the staff. Some of these were recruited and hired by the Executive Committee, some by the Secretaries themselves. Few of these lasted more than a few months. The idea of providing professional assistance to the Secretary as well as training of his potential successor is attractive, of course. In practice, however, appointees were either poorly chosen as to personal compatibility with the Secretaries and staff or were too impatient to inherit greater responsibilities. Experience of the past half century suggests that appointment of Assistant Secretaries very long in advance of their potential accession to the secretaryship may produce more problems than benefits.

The Secretaryship—Duties and Responsibilities

The Secretary/Executive Director's duties and responsibilities to the Society are manifold. Writing in 1946, Secretary Aldrich summarized his concept of the job thus: "The Secretary must be a jack-of-all-trades, as with the slogan from Ecclesiastes—'Whatsoever thy hand findest to do, do it with all thy might'" (Council Minutes, April 1946, p. 38). Later, as he was approaching the end of a long career in the position, Aldrich used four pages of fine print to list what he considered his chief duties *(Proceedings for 1955,* p. 21–25). He distinguished roughly between duties that stemmed directly from the Bylaws and those that he had had thrust upon him by Council and committee actions or other sources or merely by his own interpretations of his responsibilities. Virtually all of Aldrich's duties of 1955 are those of Secretaries today, though now for a vastly larger organization with innumerable new duties added. There is little point in listing them in detail.

The Secretary/Executive Director is the manager of a relatively large and complex business organization. He serves at the pleasure of the Council and carries out its wishes; his actions are subject to review not only by the officers, the Executive Committee, the Council, and 12,000 members but also by the Society's legal counsel, the Internal Revenue Service, the independent auditors, affiliated societies, and many other non-GSA groups. He spends and accounts for more than $2,000,000 a year of Society money and, although he must keep within an overall budget, has authority to revise internal budget items as needed. He manages a great publishing house and sells its products. He helps innumerable standing and ad hoc committees to organize themselves, hold meetings, and prepare reports. He does the same for eight divisions and six sections.

These are only a few of his duties and responsibilities. The Secretary/Executive Director does not, and cannot, do all these things himself, of course, though he must and does carry the ultimate responsibility. Nowadays, he has a staff of between 40 and 50 full- and part-time employees, each of whom is trained to perform some phase of the Society's operations. These people do most of the daily work while the Secretary/Executive Director gets the credit or blame. Selection, training, and guidance of a staff of this size and keeping it happy and productive is, itself, a management responsibility of no mean proportion.

Even though the Secretary's role is to do the Council's bidding, he is in reality the chief operations officer. With a few exceptions, he chooses the staff he needs to help carry out his responsibilities. He defines internal organization, sets pay scales, and makes salary adjustments, recommends changes in fringe benefits, and administers them after acceptance by the Council.

Perhaps more important than internal management, the Secretary may, if he wishes, have considerable influence on elective and appointive officials. For example, his advice and suggestions are commonly sought by Nominations Committees and Committees on Committees as well as by the Executive Committee and the Council. This is but natural because the Secretary is usually in a better position than most others to observe the Society's members in action and to know which ones may be best fitted to particular jobs, which ones can be depended on to deliver on their assignments, which are well meaning but dilatory, and so on.

As GSA's representative to many outside groups or as background advisor to its officers who sit on such groups, the Secretary also plays a large part in formulating Society policies and in projecting a desirable image of GSA to the rest of the scientific world.

Status of the Secretaries

From 1888 through 1961 the Society's first six Secretaries (Stevenson, Fairchild, Hovey, Berkey, Aldrich, and Betz) were elected officers and had voting seats on the Council. They served in the dual roles of Secretary to the Council and administrative officer of the headquarters office. From 1932 through 1961 the Secretaries (Berkey, Aldrich, and Betz) were also members of the Executive Committee (see Chapter 6).

Since 1962 the Secretaries have been employees of the Society rather than elected officials. They are usually appointed (hired) by the Executive Committee with Council ratification and serve at the pleasure of the Council. They sit with the Council regularly and advise it upon request, but they are not a part of it and have no vote, nor are they members of the Executive Committee. Traditionally, all of the Secretaries have been professional geologists.

The significant change in management philosophy of

classing the chief executive officer as an employee rather than an elected official was slow in coming. Removal of the Secretary from the Council was first seriously proposed by Councilor James Gilluly as early as 1943, but no action was taken by the Council either then or in 1957 when it was again urged, carefully considered, and rejected.

The thoroughly revised Bylaws adopted in 1963 removed the Secretary from Council and Executive Committee, and minor revisions of the Constitution and Certificate of Incorporation in 1967 made the Secretary legally responsible as Secretary to the Council.

During the transition period between the philosophies of Secretary/Councilor/Executive Committee member, and Secretary/employee, the old Bylaws still called for an elected Councilor to serve as Secretary to the Council and the membership. The problem was resolved by asking Joe Webb Peoples, a professor at Wesleyan University, to stand for election as Corporation Secretary. He filled this position from 1962 through 1964. From 1965 through 1967 Agnes Creagh served in this capacity.

The temporary split of corporate secretarial responsibilities from headquarters administrative responsibilities was in part an attempt to conform to the Bylaws then in effect. More important, it reflected a growing feeling that the chief administrative officer should not be a Councilor "to eliminate the conflict of interest resulting from being simultaneously an employee of the Society and a member of its board of directors" (*Annual Report for 1961*, p. 4). A clean separation between employers and employees was no doubt long overdue, for the Secretaries' seats on Council and even on the Executive Committee gave them stronger positions of power than the Society perhaps wanted them to have or than was healthy.

Depending on individual personalities, each of the later Secretaries handled his job somewhat differently, but each one was clearly a servant of the Council, there to carry out its wishes and to advise discreetly where necessary rather than to lead and speak for the entire Society as did Berkey and Aldrich for almost four decades. Which role for the secretaryship—quasi dictator or loyal servant—has been better for the Society is difficult to say. The Society has grown and prospered under both systems and has met and overcome similar obstacles under both. More has depended on personalities and interactions between them than on role definitions. In any event, no change in the currently accepted separation between policy and operations seems likely to occur in the foreseeable future.

Similar separations have been adopted by many other groups, of course, from sovereign nations to professional associations. For example, a recent announcement of a vacancy in the executive directorship of an organization similar to GSA contains the sentence: "Applicants should be able . . . *to discriminate between policy and operations* [italics added]" (*Geotimes*, v. 23, no. 9, p. 41, 1978). This

requirement is the key to successful recruiting and to performance as executive manager of any organization like GSA. But it is a two-edged sword that must be recognized by all concerned. The Executive Committee and the Council set policy; the Executive Director carries out that policy. An employee incumbent who violates the distinction will have troubles, and his tenure is likely to be short. On the other hand, the Council or Executive Committee that dips too deeply into operations, especially if impulsively, will lead to uncertainties and lowered morale or risk bad decisions.

Headquarters Staff and What It Does

Prior to receipt of the Penrose bequest, the headquarters staff was miniscule. Each of the Secretaries had a part-time amanuensis, usually the one who served him full-time in his principal employment; GSA made small allowances for the secretarial assistance. The Treasurers had similar arrangements.

To cope with vastly increased financial, publication, and other responsibilities, the paid staff grew rapidly but not extravagantly as soon as Penrose money became available.

For the next 30 years during the remainder of Secretary Berkey's regime and all of Secretary Aldrich's, the staff was remarkably stable both in size and tenure. For nearly all that period, it seldom exceeded a dozen loyal, dedicated people who tended to stay on the job for many years at relatively low salaries.

Things changed markedly shortly after Dr. Aldrich retired early in 1960. The staff grew rapidly in numbers, payrolls and individual salaries increased, and personnel turnover became a fact of life. Part of the change was unquestionably due to differences in management philosophies and methods, to rapid growth in membership, and to quantum leaps in all of the Society's activities and responsibilities. Much of it, however, arose from the changes in American life that have affected all corporations and all individuals or groups.

The Society's tax-exempt status relieves it of some of the government regulations that burden most corporations, but coping with such regulations still places demands on the staff. Similarly, while headquarters promotes efficiency with up-to-date office equipment, the use of modern computers, reproduction equipment, and so on never seems to eliminate the need for more human hands and brains.

By 1974 the staff had reached a high of 54 employees, 14 of them part-time. Successive managements and Councils had tried to restrict growth but without much success. Growth in staff numbers and almost constant turnover of employees were major factors in the lives of each Executive Secretary and in their relations with the Executive Committee and Council.

Since the major economy wave that started late in 1974, the staff has been gradually reduced, largely by attrition. By the end of 1978 the Executive Director could report that staff had been cut by 23% from the early 1975 high (*GSA News & Information,* 1978, vol. 1, no. 4, p. 50). A significant part of this cut, however, had been offset by a growing trend to contract certain Society activities to outside individuals or companies, so that the figures on staff cuts and consequent savings are not comparable.

To meet crises, numerous short-term, part-time employees are used. A recurrent crisis, for example, occurs each summer and fall when the myriad details of planning and organizing the annual meeting coincide with the more mundane but just as essential duties of preparing and distributing election ballots and dues and subscription billings for the succeeding year. In recent years, senior citizens from nearby retirement homes have proved to be rich and eager sources of help in stuffing envelopes, sorting registrations for meetings and assembling the inevitable packets of material that greet annual meeting attendees.

Staff Benefits

In modern times the Society's fringe benefits for the paid staff have been competitive with those of most employers. It was not always thus. Until the late 1930s, and in common with most of the country, there were virtually no fringe benefits such as paid annual or sick leave, insurance, or retirement annuity plans. Some departing employees received cash or other gifts, but some did not. Myra Ale was assistant to Treasurers Clark and Mathews at John Hopkins and kept most of the Society's financial books, from collection of dues to payment of bills, for nearly 40 years, from 1906 to 1944. When she retired shortly after the death of Mathews on February 8, 1944, she was given an unusually large "bonus" of half a year's salary, or $500.

The first study of possible retirement annuity plans was begun in 1937, but this did not bear fruit until 1941 when a plan was established with a private insurance company as carrier. Designed to cover all full-time employees, it was contributory, employees and Society each paying in 5% of salaries. Retirement age was set at 65, with some exceptions, and annuities were to approximate 50% of final salaries. Life insurance was available, but GSA did not contribute to this feature (Council Minutes for 1941, p. 126).

In 1951 after several years of study by a special committee, the Council voted to drop the commercial annuity plan as unsatisfactory and with inadequate benefits. Instead, it set up a self-administered plan based on actuarial principles. A special retirement fund was establishd, financed by earmarking the income from $400,000 worth of securities within the Reserve Fund. The plan was noncontributory on the part of employees and set normal retirement ages at 60 for women and 65 for men but with no mandatory top limit. It was to pay annuities of 50% of the average of the final 5-years' salary, but made no provision for beneficiaries after death of the annuitants. The plan was designated to supplement federal Social Security payments which had become applicable to retirees from small work forces like that of GSA since adoption of the earlier plan (Council Minutes for 1946, p. 80, and for 1951, p. 277).

The internally administered retirement plan soon proved cumbersome for a small business office, and employees had no equity, no flexibility, and no options; moreover, it posed great risk for the Society because the staff was small. Another committee made another and very thorough study of the problem in 1955. It consisted of Henry Aldrich, Ian Campbell, A. Rodger Denison, and E. Fred Davis, chairman. Their report was immediately accepted by the Council, and the Society joined the Teachers Insurance and Annuity Association (TIAA) in November 1955 (Council Minutes for 1955, p. 206). TIAA, a nonprofit organization, is well known in academic circles; it offers excellent retirement and insurance plans for teachers and other employees of hundreds of universities, colleges, and related groups.

The GSA-TIAA retirement plan is contributory by both employer and employee, is coordinated with Social Security, and like all TIAA plans offers immediate vesting of benefits in the employee, as well as complete portability of the employee's equity from one TIAA-member employer to another. The plan has been revised and updated through the years, and it is still the backbone of the Society's fringe-benefits program. The 1955 committee report also provided for many other benefits, such as term life insurance and group medical insurance, both paid by the Society, and adequate provision for annual and sick leave. Benefits have been revised from time to time, and the total fringe-benefit package continues to compare favorably with local industries, government, and academic institutions. The employee benefits are outlined in a handbook for employees, which is updated as needed, copies of which can be obtained from headquarters.

In short, GSA does not baby its employees, but it has for many years done its best to supply good working conditions and to pay wages and provide fringe benefits that compare favorably with other organizations in the community.

Staff Functions

This history is not the place for detailed description of the work done at headquarters on behalf of the Society membership and of geology. A reasonably complete statement of most headquarters activities appeared in a series of four articles in 1973–1974 in *The Geologist* (v. VIII and IX), the predecessor of the current *GSA News & Informa-*

tion newsletter. The staff-written stories presented the Boulder employees as individuals (with photographs) and with brief summaries of how each fitted into the manifold headquarters operations. Faces, numbers, and organizational structure have changed since 1974, but the kinds of work done were still much the same in 1980.

Through 1979, the names, titles, and employment dates of the entire staff, both full- and part-time employees, appeared annually in the *Membership Directory* and the annual *Proceedings*. Such listings were dropped in 1980. Despite changes in these lists, the basic functions performed by the staff remain essentially unchanged from year to year and from decade to decade. These functions are summarized in the following paragraphs.

Administration. The Executive Director with an Administrative Assistant and one or two others are responsible for all Society operations. Their duties fall into two main segments. One is the management of all headquarters operations—staff, housekeeping, accounting, and the like, either directly or by delegation, except as to guidance and final decisions. The other and much larger set of duties is maintenance of headquarters' relations with other societies and with all of GSA's governing or operating bodies—Executive Committee, Council, committees, sections, and divisions—and maintenance of records for all of them.

Publication Program. More than half of the headquarters staff is involved in the Society's publication program. To this total must be added the large fraction of "overhead" activities that supplement the publications function—building maintenance, administration, personnel and payroll, and others.

GSA is a great publishing house (see Chapter 10). Maintenance of the enormous flow of journals, books, maps, and other printed materials requires a relatively large staff, all of them skilled and most of them highly specialized. Through 1980 the program was under the direction of an in-house Science Editor who was primarily responsible for manuscript selection and for overall policy as defined by the Committee on Publications and the Council. Under him were the editors, proofreaders, secretaries, and production specialists who prepared an ever-increasing share of the Society's publications in camera-ready form for contract printers' presses and binderies. The internal preparation of materials for publication continues, but at the close of 1980 the internal Science Editor's position was abolished in favor of using Science Editors external to the organization. The Executive Director was assigned the additional duties of Publications Manager.

In addition to the editorial and manuscript preparation function, numerous other employees are engaged in selling and distribution of the Society's publications. Without their efforts the entire publications program would fall of its own weight, and the Society would soon become bankrupt. Sales, sales promotion, subscription fulfillment, distribution of publications to the membership—all these are just as vital to a successful publication program as the production of printed matter. So, too, accurate accounting for income and expenses for allocation of costs to individual books or to classes of publication is a vital prerequisite to determination of selling prices. This aspect of the publications program has been handled in various ways in the past with results that have bewildered successive Councils and have doubtless lost money for the Society at times. For example, for many years selling prices of books were established by applying an arbitrary markup to the production cost of an edition. The "production cost" included only the cost of printing and binding; it did not include the in-house cost of editing and other preparation. Moreover, selling prices were set on the assumption that an entire edition would be sold eventually, a risky assumption for highly technical material with a limited audience. The seemingly elaborate and onerous system of cost accounting and allocation that evolved mainly in the 1970s is better, more equitable, and more businesslike than any system that was used earlier in GSA's history.

Fiscal Services and Business Office. Until 1944, all money handling and accounting was done by the Treasurer and his secretary at their own offices rather than at GSA headquarters. From 1906 through 1943 when William B. Clark and then Edward B. Mathews were Treasurers, this meant that the Society's books were kept at The Johns Hopkins University in Baltimore. When Clark died in 1944, all financial functions were transferred to the Society's New York headquarters. Secretary Aldrich and Assistant Treasurer Beatrice Carr were nominally responsible for all income and disbursements; Mrs. Miriam Howells actually kept the books (Council Minutes for 1944, p. 19). After Henry Aldrich retired early in 1960, financial responsibilities continued to rest with the Secretaries/Executive Directors, but most of the work was delegated to staff accountants, with varying titles through the years.

Today, the Society's business transactions are so numerous and complex that at least a quarter of the total staff is engaged in one facet or another. All business functions, fiscal and others, are under the direction of a Controller, formerly called the Business Manager. As noted in the discussion of the publications program, a significant segment of the business staff is actually engaged in direct furtherance of that program. Book and map sales, subscription orders and fulfillments, printing contracts, cost accounting for each published book or series—all these and many more are just as essential to continued success of GSA's publication series as are the more generally recognized functions such as manuscript selection, editing, proofreading, and book design.

In addition to his group's part in the publications program, the Controller and his staff are responsible for drafting budgets, accounting for and safekeeping of money, maintaining personnel records and payroll (including the accounting and making payments for Social Security, various types of insurance, and income taxes). Other responsibilities include maintenance of building and grounds and coping with the mountains of mail that flow both into and out of GSA every workday. All of the services listed and many others are absolute necessities for any business as complex and far-reaching as GSA's. Despite liberal use of computers and other modern tools, a significant number of skilled people are essential to take care of headquarters' business and fiscal duties.

Meetings Department. "The purpose of the Society is the promotion of the science of geology by . . . the holding of meetings . . ." (GSA Constitution and Bylaws, Article II). A small core group, supplemented as necessary during pressure periods, constitutes the Meetings Department. While available to assist sections, divisions, and others associated with the organization in holding their meetings, the primary purpose of this group is to handle the multitudinous details associated with the GSA annual meetings. Much of their work is behind the scenes working closely with conference-location committees, local housing and conference facilities, and exhibit displayers. Technical programs are arranged by the Joint Technical Program Committee, *Abstracts with Programs* books are compiled and published by the GSA editorial staff, but provisions of space, audio-visual equipment, and all the other requirements for scores of technical, social, and other functions are handled by the Meetings Department staff.

Membership. Another small group is responsible for the care and maintenance of the records of the 12,000 Society members, from answering initial inquiries about joining the Society to finding memorialists for the deceased. Preliminary checking of credentials prior to selection of new Members or Fellows by the Membership Committee, maintaining name and address lists, preparing and mailing dues notices, and compiling the annual *Membership Directory* are only a few of the group's duties. One of the more productive and fastest-growing responsibilities is the employment service, which involves computer matching of potential employers and job applicants, plus arrangement for face-to-face interviews at GSA annual meetings.

8

Investment Management

Pre-Penrose Era

Prior to receipt of the Penrose fortune, the Society was self-supporting. Nearly all of its money came from initiation fees of $10, from the $10 yearly dues, and from subscriptions to the *Bulletin*. Aside from comparatively large printing bills for the *Bulletin*, expenses were small. There were no housing or travel costs, no legal or auditing fees, and no salaries—other than small honoraria to Secretaries and Treasurers and allowances for part-time clerical assistance to both of them. The cost of annual meetings was borne by the local host groups.

Despite this modest existence, the Society had amassed an investment portfolio of nearly $50,000 by 1930. This sum, which stands as a tribute to the unsung devotion and financial acumen of four successive Treasurers, was fed almost entirely by investment of initiation fees and life-commutations of dues. The accomplishment seems all the more remarkable when one learns that until 1930 the Treasurers and Finance Committees never had authority to *sell* any of the investments! The only changes permitted in the portfolio were investments of new funds or reinvestments of money derived from bond maturities or redemption calls. This curious restriction on investment policy, which seems to have been based only on tradition, was removed by adoption of new Bylaws December 1930. Even then the policies were cumbersome, for specific investment changes could only be authorized by the Council at its regular meetings after advice from the Treasurer and the Finance Committee. Within a couple of years after receipt of the Penrose fortune the rules were changed again, and

successive Councils routinely adopted resolutions that permitted much greater freedom on the part of the Finance Committee and Treasurer.

Penrose Era

The inheritance of the Penrose fortune in 1931 was of course a major milestone in GSA financial history; proper management of its treasury was a particular challenge. Both the provision of immediate and continuing income and preservation of the capital to meet the needs of the future had to be considered.

The record of money management has been meticulously reported regularly to the membership. Annual reports show the essential facts as to income and outgo and the state of the investment portfolio, all supported by audits from independent auditing firms. (Prior to 1931, audits were performed by committees drawn from the membership.) For most years, too, security holdings in all of the investment accounts have been published. Nevertheless, awareness of the state of the Society's finances has been relatively low among the members. Even though the dollars-and-cents details lack appeal to most, it seems essential to include a summary of GSA's financial history here, for on success or failure of the money-management program rest all of the Society's activities—past, present, and future.

As one of its first acts in adjusting the Society to its new wealth, the Council appointed a special Advisory Finance Committee consisting of H. Foster Bain, chairman; Ernest Howe; and George Otis Smith. The committee's three-page report, which paralleled and supplemented that of the Advisory Committee on Policies and Projects (see Chapter 4), was presented to the Council on April 4, 1932 (Council Minutes for 1932, p. 169–171). Almost all of its recommendations were adopted, the Bylaws were changed to conform, and the resultant policies and practices were followed for many years with but little change.

Principal recommendations in the report are listed here:

1. Maintain principal of Penrose bequest jealously intact with only the income available for spending.

2. Restrict investments to securities approved as legal investments for savings banks in Massachusetts or New York.

3. Add realized capital gains to principal.

4. Start a reserve fund to receive 10% of current income until otherwise ordered by Council; to be used to restore the principal in case of depreciation. Use the Society's pre-Penrose capital assets as foundation for the reserve fund.

5. Diversify investments with preference for securities with wide and ready market.

6. Establish a Finance Committee of three (or more) with responsibility for investing funds. Committee should be elected by the membership and should not serve on Council.

7. Treasurer, to be appointed by the Council, to be formal custodian and disbursing officer for all funds; to advise the Finance Committee; but to buy and sell securities only as directed by that committee.

8. Keep all securities in custodian account with a bank or trust company.

9. Appoint an Assistant Treasurer (at Council's discretion).

10. Bond fiduciary officers and employees at Society expense.

11. Centralize management of the Society and funds in one place under administrative authority of the Secretary.

12. Employ a competent firm of certified public accountants.

13. Amend Bylaws as necessary to fit the new policies.

The Ultraconservative Period, 1932–1947

For many years it has been the fashion among those involved in the Society's finances to poke wry fun at the early financial managers. Specifically, those managers are accused of shortsightedness and poor judgment by investing only in high-grade, low-income bonds issued by the federal government, railroads, and public utilities. The accusation is correct, but viewed from a perspective half a century later, the implications of bad management are scarcely supported by the facts and most certainly the blame, if any, has been consistently aimed at the wrong targets.

For the entire period of 1932 through 1948, the Society's lawyer and the laws he had to interpret had far more effect on the basic investment policy than had any of the specific investment decisions of the Finance Committee, the Treasurer, or the Assistant Treasurer.

From the very first, Allen R. Memhard, the Society's attorney, took the firm position that the Penrose Endowment Fund must be invested under the New York State laws governing investment of trust funds. This position, incidentally, was identical with that of the special Advisory Finance Committee. Memhard was a sound, experienced, and conservative lawyer who had the Society's best interest at heart. Neither he nor the equally sound members of the Advisory Committee should be taken to task for embarking the Society on an ultraconservative fiscal course. But if blame there is, it should be leveled at the laws of the time and the interpretations of them by the attorney and the Advisory Committee rather than at successive Councils, Treasurers, and Finance Committees who followed that advice.

Memhard's interpretation, whether wholly correct or not, was the guiding principle for GSA's investments for 20 years; it automatically limited those investments to issues that were "legal" for trust funds—which meant high-grade government, railroad, and utility bonds.

The investment history for the first 20 years of the Penrose fortune is admirably recorded by Assistant Treasurer Beatrice Carr in a report to the 1951 Council entitled "The Penrose Bequest: A Calvalcade of Twenty Years" (Council Minutes, 1951, p. 287–386). For various reasons this report was not published in the *Proceedings,* but it exists in the headquarters files and should be studied by anyone who is ever tempted to look down on the quality of early money management.

Looking back, it is difficult to see how any other investment program might have served the Society better during its first 20 years of wealth. The inheritance itself consisted almost entirely of safe but low-income, tax-exempt municipal bonds and cash. All of this had to be converted to "legal" investments in the midst of a great depression. This depression accompanied by devaluation of gold and by other unsettling events was quickly followed by still another depression and soon thereafter by a devastating war and its aftermath. Through it all, the total market value of the Society's portfolio fluctuated widely; for several years it even dropped perilously close to the original value of the Penrose gift. In good times or bad, however, no investments were sold at a loss (even though unrealized capital gains, hence capital growth, were relatively small). Even more important, income sufficient to meet the ever-growing demands of the Society was maintained even in the depths of the depression. Whether a more growth-oriented investment program would have yielded any better overall results in terms of the Society's long-range welfare will never be known, but considering the legal interpretations to which they were committed, the Society's financial managers did extremely well.

The 1948–1953 Upheaval

The period 1948–1953 was one of great unrest and profound changes in the financial philosophy and money management of the Society. Most of the unrest reflected massive changes that marked the post–World War II years, particularly changes in investment climate and philosophy in the financial community. The councilors and other Society members began to raise serious questions as to the wisdom of the investment policies that had served GSA well since the receipt of the Penrose bequest. Attorney Memhard, who was responsible for much of the conservative policy, resigned when the Society sought another legal opinion. The Treasurer and Chairman of the Finance Committee, W. O. Hotchkiss, left soon after, feeling that he, too, had been discredited. Miss Beatrice Carr, Assistant Treasurer, clashed with the new head of the Finance Committee, and only an overwhelming vote of confidence from the Council persuaded her to stay on until 1953. She retired then only for reasons of health and age, but not before she had a chance to help set the Society on a new and more liberal investment course.

As early as 1948, the new attorney, Hugh Satterlee, had raised the interesting question as to whether the Society must consider itself controlled by state and federal laws as a trust, or under the less restrictive laws that applied to corporations. He pointed out that Penrose had insisted on incorporation of GSA in 1928 and had specifically mentioned the corporate status in his will. Whether Mr. Satterlee was right or wrong in this opinion, the first real change in investment policy waited until 1950. In that year a new law was adopted by the legislature (New York State Laws of 1950, Chapter 464). This amendment relaxed many of the restrictions that had plagued trust fund managers for many years. From that time forward, trust and bank funds could be invested in equities as well as in "safe" or "legal" bonds. The Finance Committee welcomed the relaxation of restrictions and immediately embarked on a new kind of investment program—one that involved a wise mix of investments in both bonds and stocks. The new—to GSA—philosophy has been followed ever since with variations to fit changing economic conditions and with gratifying results.

Credit must go to J. Edward Hoffmeister for guiding the Society through the 1948–1952 financial embroilment and for setting it firmly on what he called the "period of growth and relative tranquility (1953–1980)." Then Dean of the College of Arts and Sciences at the University of Rochester, Hoffmeister was persuaded in 1947 by President James Gilluly to accept the job as Society Treasurer. Blessed with common sense as well as investment knowledge, he was a good choice. He smoothed the troubled waters, found an outstanding and understanding tax lawyer, Hugh Satterlee, and persuaded his good friend Hulbert

W. Tripp to take Beatrice Carr's place as investment adviser when she retired. Tripp was associated treasurer for the University of Rochester and advised the Society brilliantly for 10 years, receiving only a small honorarium. Hoffmeister was Treasurer for 15 years and also chairman of the Investments Committee for much of that period. When he finally retired from the committee in 1968, he received a glowing tribute and an impressive certificate from his colleagues at the Annual Meeting banquet in Mexico City.

The Treasurers and Investments Committee members who followed Hoffmeister were no less competent or loyal than he. They all did superb jobs in managing the Society's wealth, but their tasks were easier because of the principles Hoffmeister had established during the painful transition of 1948–1952 from ultraconservative to restrainedly liberal policies.

The Period of Growth and Relative Tranquility, 1953–1980

The modern-day investment policy of the Society is well told by Goldstein (1973), together with a brief update by Heroy and Goldstein (1976). In reality, the policy has evolved gradually from 1953 onward with less and less Council-imposed restrictions on the proportion of common stocks in the portfolio and with increasing emphasis on capital growth. Both of these trends mainly reflect changes in the investment climate and in the tax laws.

The overall results of the Society's investment program from 1931 through 1980 are shown in the accompanying table. This is based on published annual reports to the Council and membership from Treasurers, Finance (Investments) Committees, and the independent auditors.

Tabulation of totals of market values and of investment incomes of the entire investment portfolio was chosen as the only meaningful record of investment management through the years. Book values of investments are important to auditors and others, but they have no real bearing on the wealth of the Society. Similarly, figures on the separate funds that make up the overall portfolio are of great importance in management of those funds and use of their proceeds, but to detail the records of each fund would be difficult to accomplish and the results would be more confusing than instructive to any but the most financially sophisticated and interested readers.

In summary, the table shows that the value of the investment portfolio by 1980 had more than doubled from that of the original Penrose bequest. (Any attempt to record growth of the endowment in constant dollars would almost certainly prove discouraging.) At the end of 1980 total market value was $9,591,172, plus more than $1 million in land and improvements. Within the period 1931–1980, however, total values ranged between wide limits, largely in response to national and international economic

MARKET VALUE AND ANNUAL INVESTMENT INCOME
(ALL INVESTMENTS, INCLUDING CASH)

Year	Market value (000 omitted)	Investment income[1] (000 omitted)	Year	Market value (000 omitted)	Investment income[1] (000 omitted)
1931	. .	100[2]	1968	10,579[6]	420
1932	$3,884[3]	146	1969	9,782	466
1933	4,084	162	1970	9,729	415
1934	4,221	163	1971	10,295	426
1935	4,409	174	1972	11,058[7]	407 + 287 gains[8]
1936	4,570	178	1973	9,101	411 + 243 gains
1937	4,785	186	1974	6,817	373 − 371 losses
1938	4,842	198	1975	7,679	365 − 354 losses
1939	4,898	198	1976	8,450	407 + 64 gains
1940	3,899	200	1977	7,999	439 + 135 capital and option gains
1941	3,908	210	1978	8,003	504 − 26 losses on options and + 99 capital gains
1942	3,958	212			
1943	4,351	214			
1944	5,544	213	1979	8,631	566 + 241 capital and option gains
1945	6,117	228			
1946	5,784	225	1980	9,591	689 + 268 capital gains
1947	5,529	218			
1948	5,543	182			
1949	5,297	213			
1950	5,500	183			
1951	5,500	200			
1952	5,734	222[4]			
1953	5,639	227			
1954	6,370	232			
1955	7,106	241			
1956	7,740	262			
1957	7,456	274			
1958	7,272	269			
1959	8,028	270			
1960	7,368	277			
1961	8,574	275			
1962	8,266[5]	312			
1963	8,815	331			
1964	9,851	353			
1965	10,051	381			
1966	9,105	393			
1967	9,753	402			

[1]Does not include income from dues, publications sales, and other noninvestment sources.

[2]Income earned by Penrose's holdings after his death but before settlement of estate; used in support of Sixteenth International Geologic Congress.

[3]Original value, Penrose bequest.

[4]Completed transition from total investment in "safe" bonds to mix of stocks and bonds.

[5]Plus investment of $275,000 in headquarters building, its rehabilitation, and furnishing, New York City; when building was sold in 1968 for a small profit, proceeds were set aside for future purchase of Boulder, Colorado, building.

[6]Began relatively heavy investment in growth stocks.

[7]Plus 1971–1972 investment of about $1,000,000 in Boulder headquarters land, buildings, and equipment.

[8]Began to consider realized gains and losses as part of Society income available for current operations.

conditions, but in part to varying investment philosophies.

Gratifying as it may be in itself, the overall growth of the Society's capital is perhaps less important than the size and continuity of income that has been earned by that capital. From inception in 1931 through 1980, income has totaled more than $12 million. Virtually all of this money, three times more than Penrose's original gift, has been used directly in carrying out the Society's objectives. For many years investment income was by far the largest single source of operational funds. More recently, however, the proportion of income from investments has gradually declined as other sources such as dues from an ever-increasing mem-

bership and publication sales have zoomed upward. In 1969, the investment income dropped below 50% of total income for the first time. This trend might have continued in any event, but it became mandatory when the Tax Reform Act of 1969 (Public Law 91-172) required that no more than 33 1/3% of total income could come from investments if an organization like GSA were to retain its tax-free status as a nonprivate (public) foundation. This rule placed but one more difficulty in the path of the Society money managers, but it was surmounted as were many others.

In addition to income from dividends and interest,

THE GEOLOGICAL SOCIETY of AMERICA
SOURCE & APPLICATION of FUNDS ✳ 1978

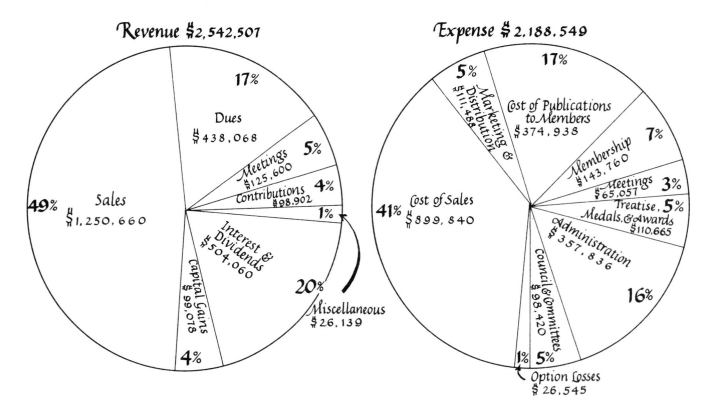

probably a million or more dollars of net realized capital gains have come to the Society from sales of securities that had appreciated in value. The total cannot be easily calculated because until the last few years realized gains and losses have appeared in financial statements only sporadically. More important, realized capital gains were until recently always ploughed back in investments of new securities for the portfolio, hence were not considered spendable income.

In 1973 the Society was advised, *again by its legal counsel,* that there were no legal barriers to use of realized capital gains for operational purposes. This advice was soon put in practice even though it carried with it the inevitable corollary that realized capital *losses* must also be considered in reports of operating results.

Separate Funds

Ever since receipt of the Penrose bequest, the Society's money has been held and accounted for in more than one fund. As many as 12 distinct funds have been maintained at times, each differing from the others in size, source, purpose, and investment objective. The largest by far has always been the Penrose Endowment Fund itself, which has seldom, if ever, held less than three-quarters of the Society's

total capital. With a few exceptions, all of the funds represent parts of the Penrose fortune, for they were established by setting aside segments of the Penrose capital or income to provide for specific purposes. The exceptions are a few relatively small funds that were financed by other sources, such as the various medals and awards funds, and the larger Current Operating Fund, which is fed by dues and sales receipts as well as by investment income. The total of all these exceptions is so small that the Society's overall investment history is really that of the management of the Penrose bequest.

The separate funds have changed through the years in nomenclature as well as in purpose, but the management of numerous funds, each with its own portfolio and investment objective, has posed problems for successive Investment Committees and for the accountants. In mid-January 1978, the Committee on Investments wisely decided to pool the 12 existing funds for investment purposes but not for internal accounting, into four groups: (1) the Endowment Fund, comprising the bulk of the Society's invested capital; (2) the Reserve Fund, for long-term and periodic needs such as major repairs and employee retirement; (3) the Current Fund, which provides cash to support the Society's everyday operations; and (4) the Medal and Awards Fund. This treatment has greatly simplified the tasks of the money

GSA FINANCIAL MANAGERS 1888-1979

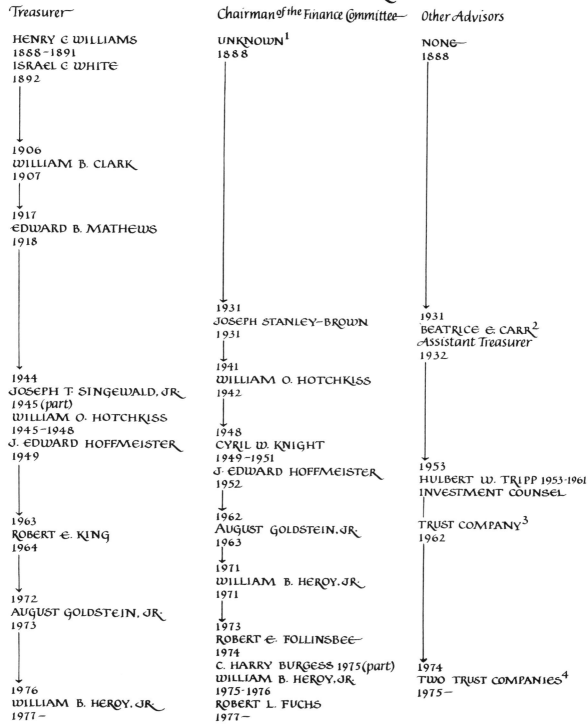

Treasurer	Chairman of the Finance Committee	Other Advisors
HENRY C WILLIAMS 1888-1891	UNKNOWN[1] 1888	NONE 1888
ISRAEL C WHITE 1892		
1906		
WILLIAM B. CLARK 1907		
1917		
EDWARD B. MATHEWS 1918		
	1931 JOSEPH STANLEY-BROWN 1931	1931 BEATRICE E. CARR[2] Assistant Treasurer 1932
1944 JOSEPH T. SINGEWALD, JR. 1945 (part)	1941 WILLIAM O. HOTCHKISS 1942	
WILLIAM O. HOTCHKISS 1945-1948	1948 CYRIL W. KNIGHT 1949-1951	
J. EDWARD HOFFMEISTER 1949	J. EDWARD HOFFMEISTER 1952	1953 HULBERT W. TRIPP 1953-1961 INVESTMENT COUNSEL
1963 ROBERT E. KING 1964	1962 AUGUST GOLDSTEIN, JR. 1963	TRUST COMPANY[3] 1962
	1971 WILLIAM B. HEROY, JR. 1971	
1972 AUGUST GOLDSTEIN, JR. 1973	1973 ROBERT E. FOLLINSBEE 1974	
	C. HARRY BURGESS 1975(part) WILLIAM B. HEROY, JR. 1975-1976	1974 TWO TRUST COMPANIES[4] 1975—
1976 WILLIAM B. HEROY, JR. 1977—	ROBERT L. FUCHS 1977—	

[1] During the latter part of the period 1888-1930, small finance committees were appointed anually, but names of their chairmen, if any, do not appear in the records.

[2] Miss Carr was aided by a trust company, 1949-1951, custodial and advisory, non-discretionary

[3] Custodial and advisory, non-discretionary

[4] Portfolio divided between two companies, custodial and advisory, but with discretionary management in part.

Treasurers

Henry S. Williams, 1847–1918; Treasurer, 1889–1891.

Israel C. White, 1848–1927; Treasurer, 1892–1906; President, 1920.

William B. Clark, 1860–1917; Treasurer, 1907–1917.

Edward B. Mathews, 1869–1944; Treasurer, 1918–1944.

Joseph T. Singewald, Jr., 1884–1963; Acting Treasurer, 1945.

managers (Report of the Committee on Investments: *Geology,* 1978, v. 6, no. 5, p. 281–282).

Of the separate funds, the only one that requires further discussion here is the Reserve Fund. Known by several different names at times, this fund was established at the same time as the Penrose Endowment Fund itself. Its primary purpose was to serve as a cushion for the Penrose Fund and to insure protection of the corpus. Until 1946 the Reserve Fund regularly received 10% of all Penrose Fund income. This accretion was stopped by Council action and a Bylaws change, and the only additions to the fund for some years came from reinvestment of self-generated income and capital gains within the Reserve Fund itself. More recently the Reserve Fund was fed by addition of capital gains (rather than income) realized from the Penrose Endowment Fund and from itself. In addition to its traditional use in protection of the endowment, the Reserve

Fund had been called on as a ready source of money for large and unusual expenditures, such as the purchases of headquarters buildings first in New York and later in Boulder. During the recession of 1973–1974, the Society's needs were so great as compared to its income that the entire Reserve Fund was exhausted to bridge the gap until the budget and spending pattern could be readjusted. Removal of the Reserve Fund as a security blanket alarmed conservatives in management and membership, but immediate steps were taken to rebuild it by regular transfers of net realized gains from investments.

During the late 1970s the concept of the Reserve Fund was again changed. All income from the Penrose Endowment Fund began to flow directly into the Reserve Fund, and then it flowed into the Current Fund as needed. Unlike the Endowment Fund which remained largely in long-term investments, the Reserve Fund was to be in short-term,

Treasurers

William O. Hotchkiss, 1878–1954; Treasurer, 1945–1948.

J. Edward Hoffmeister, 1899– ; Treasurer, 1949–1963.

Robert E. King, 1906– ; Treasurer, 1964–1972.

August Goldstein, Jr., 1920– ; Treasurer, 1973–1976.

William B. Heroy, Jr., 1915– ; Treasurer, 1977– .

highly liquid investments. In 1980 a limit of $500,000 was placed on this fund, with excess monies to be returned to the Penrose Endowment Fund (Council Minutes for Nov. 15–16, 1980 meeting, p. 4, 14).

The Money Managers

Aside from the Council's overall responsibility, actual management of the Society's money has always been entrusted primarily to its Treasurer and the Finance Committee (now Committee on Investments). Divisions of responsibilities between the Treasurer and the committee have varied somewhat through the years, and there has been much back and forth movement of individuals, but the two entities have interacted well together and have provided a strong cohesive force in managing the Society's funds for the common good. All Treasurers and Finance Commit-

teemen have been outstanding geologists drawn from the membership, but nearly all of them have also been men who were accustomed to handling of wealth, either their own, usually made from the ground, or in management of family or other investment funds. To their combined wisdom, experience, devotion, and volunteered time, the Society is forever indebted for the continued health, vigor, and productivity of its investment program.

The names and terms of office of the Treasurers and the Chairmen of the Finance Committees are shown in the accompanying table. Other members and conferees of successive Finance Committees are not listed, but they constitute a sizeable group of equally astute and loyal financial advisors. Noteworthy in the table is the brevity of the list of money managers over a span of 90 years and the long lengths of service of most of them, facts that have contributed greatly to stability and continuity in the Society's investment philosophy. Because the Treasurer has always been an elected member of Council and because several Treasurers have also chaired the Finance Committee for long periods, they have, moreover, provided more continuity for the Council than could any groups of Councilors or officers elected for three-year terms.

In addition to the volunteer work of the Treasurers and Finance Committees, the Society has habitually depended heavily on paid financial advisors. Whether on the Society staff or outside, these advisors have influenced the investment program greatly, both as to general policy and as to specific investment decisions. It must be emphasized, however, that the Finance Committees under powers delegated by the Council have until very recently invariably reserved the rights of all final decisions to themselves. This is as it should be, for the members of the Finance Committees have uniformly been good financial advisors themselves. In fact, most of the paid advisors have soon learned that they gain as much knowledge from the committees as they give to them.

When Joseph Stanley-Brown assumed the Society's financial guidance in 1931, his first step was to persuade a former Wall Street associate, Miss Beatrice E. Carr, to return from retirement in her native England and to become the Assistant Treasurer, a new full-time paid position. Miss Carr had been long recognized on Wall Street as an outstanding investment advisor, with the title of Chief Statistician and Investment Analyst for one of the larger brokerage houses. As such, she retained innumerable friends in the financial community who were helpful to her in her work for GSA.

Miss Carr's first duty for GSA was to take care of all the details of acceptance of the Penrose bequest including the inventory and valuation of its securities. From then till her retirement in 1953, she devoted all her financial wisdom and all her time to day-by-day management of the GSA investment portfolio. The story of her stewardship is told modestly but clearly in her "Cavalcade of Twenty Years" mentioned on another page. She retired in 1953 and died in 1957 at the age of 77.

Miss Carr's place was taken by Hulbert W. Tripp, who served as Investment Counsel on a modest honorarium from May 1953 through 1961. As the Associate Treasurer for Rochester University he was responsible for its large investment portfolio. A further advantage to the Society was that Tripp and Treasurer Hoffmeister were friends and associates on the Rochester campus, which made for easy and frequent liaison.

When Mr. Tripp resigned as Investment Advisor in 1961, the Finance Committee turned to strong trust companies for advice. This has been done only once before, during the investments turmoil of the 1948–1952 period. Results then had been less than satisfactory to the committee, and they were not much better during 1962 and 1963. From 1964 forward, after a change in advisors, the relationship between the Society and its advisors was excellent. Several companies were appointed in succession; each acted as custodian of the Society's securities as well as providing investment advice and management on a nondiscretionary basis. Beginning in 1976 the portfolio was split between two advisory firms for greater safety. One of these was given discretionary authority over some of the portfolio. Also in 1976, the committee entered the options market for the first time, giving authority to one of the advisors for writing call options. Over the next several years this arrangement proved to be only moderately successful, and it was soon abandoned.

The Society's record of financial management over a period of nearly 50 years is an enviable one. The Treasurers and Finance Committees have coped successfully with the usual ups and downs of security prices that result from economic conditions as well as from market psychology. Their investment decisions have also been influenced by legal opinions as to the categories of investments that were permissible and by restrictions on the accumulation of too much capital or of earned income. Either of these might easily affect the Society's tax-exempt status, as well as the tax status of those who pay dues or make contributions to it. Through it all, the money managers have trod a clear path between two principal objectives—to produce spendable income for uses of the present generation and to preserve adequate income-producing capital for future generations of geologists.

9

The Society's Homes

"He [R.A.F. Penrose, Jr.] once remarked that no society could impress even its own members with its dignity, without having a home of its own" (Berkey, 1936). Penrose made this remark several years before his death when he was distressed by the sight of a great Society's sole headquarters cramped into the corner of a professor's office. Penrose's generous will and the generosity of Columbia University, made it possible for the Society to acquire and staff a fine headquarters shortly after his death, but not until 1972, more than 40 years later, did it finally build a real home of its own, designed to its needs, and with the dignity that Penrose so much desired. During the intervening years, three headquarters buildings were occupied. The first was a brownstone residence borrowed from a university, the second a renovated tenement owned by the Society, and the third a rented part of a commercial office building.

During its first 42 years, the Society's headquarters were in the offices of its Secretaries with no specifically assigned space or equipment for their Society activities. Headquarters, therefore, were successively at the University of the City of New York (1888–1890), Rochester University at Rochester, New York (1891–1906), American Museum of Natural History, New York (1907–1922), and Columbia University, also in New York City (1923–1932).

In January 1930 and in response to Penrose's urging and financial assistance, Secretary Berkey arranged with Columbia University for assignment of a two-room suite in Schermerhorn Hall for his Society business. The rooms were furnished by the University, but according to Assistant Secretary Miriam F. Howells (letter of November 12, 1977), the GSA property at that time consisted of her typewriter, a hand-operated Addressograph, card and letter files, and a collection of officers' photographs which she had assembled.

GSA House, 419 W. 117th Street, New York City

In mid-1931, shortly after Penrose's death, Columbia offered GSA the use of one of its residential buildings on the edge of the campus. This building, at 419 W. 117th Street, was to become known as the Geological Society of America House and to serve as the Society headquarters for many years. Terms were generous—the loan was indefinite as to time, rent was free, and GSA was to pay for repairs, labor costs of maintenance, and utilities. Purchasing of office supplies was done through the University at substantial savings. The Council gratefully accepted the University's offer in September 1931 and began the necessary alterations. The house was furnished almost entirely with items from Penrose's personal office and library in Philadelphia. GSA moved in during May 1932, and the Council first met in GSA House on June 25, 1932.

Like many other New York City buildings of the period, GSA House was handsomely built of brownstone (Triassic sandstone). Railings and ornaments leading to the door were white limestone, replete with a variety of dwarf fossils. Steps and walk were slate (per Agnes Creagh, letter of May 19, 1979).

The West 117th Street house had five stories. It also had a basement for storage of Society publications. When GSA outgrew the storage space available, Columbia University again came to the rescue and donated for many years commodious space in its air-conditioned library.

Entrance way to GSA House, the Society's first home of its own, 419 W. 117th Street, New York City.

space in the Treasurer's office and home at The Johns Hopkins University in Baltimore until Treasurer Mathews died in 1944 and the fiscal records and responsibilities were transferred to Society headquarters.

Though Columbia's loan of the West 117th Street house was nominally rent free, it actually cost the Society a significant amount of money through its 30 years of occupancy. Initial alterations cost about $8,000, and further alterations, maintenance, and operating expense, such as heat and light, cost additional sums annually. More important than these expenses, perhaps, were voluntary donations to Columbia in lieu of rent. In 1940, and in response to repeated urging by Secretary Berkey, the Society donated $10,000 to the University's research fund. This and regular annual donations of $1,500 to $3,500 thereafter were always gratefully acknowledged by the Columbia officers who wrote that while the University neither asked nor expected any rent, the gifts were appreciated and would be turned over to the geology department in support of research in GSA's own science.

The first floor was used for reception and for production and mailing facilities; the second was set aside as a Penrose Memorial "Room" but was also the Secretary's office. The third and fourth floors were the main working offices; the Council met on the third. The top floor had a double bedroom and two single rooms, with baths, for visiting Councilors and other members, including many foreign visitors who preferred the congenial (and free) atmosphere of GSA House to downtown hotels. Up to about 200 visitors per year enjoyed the Society's hospitality. GSA had no dining facilities, but visitors were entertained (at Society expense) at the nearby Columbia Faculty Club.

According to Mrs. Howells (letter of March 25, 1979), the decision to provide guest rooms on the top floor was originally made at Columbia's request to forestall future pressures on the University from some other group for permission to fill the unused space. The Society was asked not to compete with nearby hotels but to limit guest room use to those on GSA business and to special foreign guests.

In addition to the office space and facilities generously supplied by Columbia University, the Society was for many years supplied office space for the *Bibliography* staff in the U.S. Geological Survey library in Washington. It also had

GSA's second New York City home at 231 E. 46th Street.

Second New York Home, 231 East 46th Street

Toward the end of Henry Aldrich's regime as Secretary, it began to become apparent that Columbia's loan of a headquarters building was drawing to a close. Kindly disposed to GSA as it still was, the University planned to raze the entire block of buildings that included GSA House. The University also served notice that it would soon need the library space that had been lent for storage of Society publications. This was by no means the first warning received by the Society. For years, both Secretaries Berkey and Aldrich had periodically reminded the Council that Columbia would one day have to reclaim its building.

A committee was appointed in 1958 to advise on the site for a new home, but the actual move to new quarters did not take place until five years later. The first Headquarters Committee, obviously chosen with broad geographic representation in mind, consisted of the following: President Raymond C. Moore, Lawrence, Kansas; Marland P. Billings, Boston, Massachusetts; Henry R. Aldrich, New York, New York; Byron N. Cooper, Washington, D.C.; Edwin D. McKee, Denver, Colorado; Ian Campbell, Pasadena, California; and J. M. Weller, Chicago, Illinois.

The committee was reappointed in 1959, but Dr. Moore was replaced by incoming President Billings, and Hollis D. Hedberg of Princeton was added to the roster.

The chief question considered by the Headquarters Committee was whether the Society should remain in New York or move to some other city. This freedom of choice was in itself a milestone in GSA history, for it had always been assumed that it would remain in New York City indefinitely.

The committee made no very thorough studies, but it did carefully consider most of the arguments—legal,

Entrance hall, 419 W. 117th Street, New York City. Oak table was presented by George F. Kunz, one of the founders.

Penrose Room and Secretary's office, 419 W. 117th Street, New York City, showing
Penrose's personal desk and other office furnishings.

human, and financial—for and against remaining in New York. It also considered the arguments, pro and con, for centering the Society in any one of a large number of other American cities. The committee finally reported to the Council in March 1959 that it was almost equally divided. A slightly larger group was in favor of staying in New York if some continuing arrangement could be made with Columbia, but if not, felt that the Society should consider moving to another city, "presumably" Denver. The smaller group voted to move elsewhere immediately, to Chicago, St. Louis, or Denver. Despite the committee's slight weight of opinion toward a move, the Council decided against leaving New York for the time being.

Secretary Aldrich, and his successor, Frederick Betz, Jr., performed most of the search for a new home in New York with help from members of the Executive Committee and Council. First efforts centered on the possibilities of obtaining space from Columbia University. The University was cooperative and friendly and did its best to help solve the problem. Numerous choices were offered from among its widespread properties. Several houses near the campus, at least one down-at-the-heels office building in midtown Manhattan, and even an attractive estate on Long Island were examined, but none met the needs of the Society. Moreover, all would have been relatively costly, for the terms now involved rental payments plus alterations rather than the loan of former years.

The search was then widened beyond the University. Among the more promising candidates that emerged were rental space in a new science center proposed by the New York Academy of Sciences, the then-building United Engineering Center which was expected to be ready for occupancy in late 1961, and the American Institute of Physics (AIP) building on 45th Street near 1st Avenue in New

York. From these, the Council chose the AIP space and a 5-year lease was signed in September 1961. Various frictions and doubts between the two groups developed soon after, however, and GSA's Executive Committee cancelled the lease during the ensuing year, luckily before any actual move from the Columbia campus had been initiated.

The renewed search for a home was more productive than earlier efforts had been. At its November 1962 meet, the Council authorized purchase of a building at 231 East 46th Street. The new home was bought January 23, 1963, for $174,000 plus about $75,000 in alterations and furnishings, or a total of $250,000. It was occupied soon after purchase, and the Executive Committee first met in it July 11, 1963. When the Society moved to Colorado 4 years later, the East 46th Street house was sold for a modest profit to the Venezuelan Embassy to the United Nations.

The responsibilities of house hunting, real estate purchases, engagement of an architect, and supervision of heating and air-conditioning contractors were all new to the then Executive Director, Agnes Creagh, whose professional life had been in the fields of geology and editing. She assumed them cheerfully, however, and did a superb job, but the unaccustomed strains added to supervising the physical move to new quarters and maintaining Society operations undoubtedly hastened her decision to resign.

The East 46th Street house was comparable in size and shape to the one at Columbia—five stories, 26 feet wide, and 100 feet long. Built of brick, it was then about 50 years old, had been converted from a tenement about 1930, and modernized in 1954. It actually had less usable space than the old home because the basement was needed for utilities and the top floor had to remain vacant because of fire regulations.

Space for storage of back-stocked publications posed no problem as arrangements were made with the Society's printer, Lane Press of Burlington, Vermont, to handle storage and mailing (an arrangement that was still in effect at the end of 1980. Provision of sleeping accommodations for

A part of the Penrose library, 419 W. 117th Street, New York City. This room on the fourth floor was used by Assistant Treasurer Beatrice Carr.

visitors was also discontinued, which added to the usable space (although one small bedroom was retained for occasional emergency use). Council and other official visitors were fed at nearby restaurants.

The East 46th Street house in New York was adequate for the Society's requirements, although it continued to generate needs for additional repairs long after occupancy by GSA. It was, moreover, attractive and dignified both inside and out. Close to the United Nations complex and within walking distance of Grand Central Station, it was in what had become one of New York City's better neighborhoods, surrounded by modern, high class apartment and office buildings. Had GSA remained in New York, the first house it had ever owned would doubtless have served it for many years, although the Society's growth in membership and the consequent demand for more headquarters services would have eventually placed strains on the available office space.

The Move West

The decision to move westward and away from New York was made far more abruptly than most of the Society's major policy decisions. True, it had been foreshadowed by casual suggestions from time to time and by the split decision of the Headquarters Committee report to Council in 1959.

Barely 3 years after the East 46th Street house was occupied, in April 1966, serious Council discussion of the desirability of a move first surfaced. Much of the pressure to move was instigated by Secretary Raymond C. Becker who, besides being a westerner himself, was more aware of the difficulties of operating a Society in New York City than were the Councilors who visited headquarters only once a year. He realized, too, that less than a score of GSA members, aside from Councilors and committee members, ever visited headquarters in any one year. This was in contrast to earlier years when many members had used GSA's home as a gathering place. Less than 18 months later, in September 1967, not only had the decision to move away from New York been made but also a site had been chosen, an office rented, and the physical move of headquarters accomplished.

After the April 1966 Council meeting, the membership was polled. Although the 40% response was much greater than usual, the results were predictably inconclusive—relatively few GSA members cared very much about the Society's location.

An ad hoc Committee on Headquarters Location was appointed shortly after the November 1966 meeting of Council (Council Minutes for 1966, p. 17). It consisted of Ian Campbell, Stephen E. Clabaugh, and Luna B. Leopold, Chairman. The committee's report was presented to the Council in May 1967. All three committee members con-

tributed heavily to it, as did Executive Secretary Becker and many other Society members, but it bears the unmistakable imprint of Ian Campbell in its forthright objectivity.

The ad hoc committee began work on the assumption that the Society would move away from New York City. Among the arguments for the move were these: increasing difficulty in finding and retaining adequate staff, rapidly increasing living and operational costs, deterioration in city safety and other services, never-ending expenditures for repairs and further alterations of the 231 East 46th Street building, and hopes that a move would aid in the search for a full-time Editor. Arguments against a move included fears that the Society as a New York corporation (and legally still a ward of the Orphans Court) might not be allowed by the courts to move out of the state and to take its fortune with it, doubts as to whether the Society headquarters could serve geology better away from "the education-oriented science center of the country," potential loss of much of the existing staff, and even objections to leaving the cultural advantages of New York in favor of the "cultural wasteland" of the western United States.

The ad hoc committee collected data on 16 American cities, from Albany to San Francisco, by correspondence with selected geologists and by study of material supplied by Chambers of Commerce and like groups. Factors thought to be important to the headquarters location were given relative priority values, then weighted logically. The study was made as objectively as possible, although subjective thinking surely played a part. The factors considered, and weights assigned, were as follows:

Cost of living	1
Availability of personnel	4
Airplane and airport service	4
Bus and city transport	1
Proximity to universities	1
Proximity to other geological centers	1
Climate	2
City size	2
Availability and cost of rental office space	3
Living conditions	2
Recreation	1

The committee winnowed its list of 16 cities down to five, then narrowed to the Denver area as its firm recommendation. After extended discussion of the committee's report, the Council voted, by a rare split decision, on May 7, 1967, that "the operational headquarters of GSA be moved to the Denver area at the earliest possible time and that the Society also establish a principal office in the city, county, and state of New York with such personnel and documents as may be required by law." Details of timing and of choosing a specific place were left to the Executive Secretary, moving expenses of $50,000 were allocated from the General Reserve Fund, and the Society's lawyer, Rollin

Browne, offered use of his office as the statutory New York headquarters.

No sooner had the Council acted favorably on the recommendation than planning for the move got under way. Based on a firsthand study of all the possible locales as well as on his many years of residence in Denver and its environs, the Executive Secretary settled on Boulder as the ideal location. A relatively small, self-contained city, Boulder was still within the Denver metropolitan area and only 30 miles from its heart. Probably the most compelling reasons for choosing Boulder rather than a site closer to Denver proper were proximity to the University of Colorado, the ready source of superior potential employees at all levels, and physical separation from potentially undue influence of the burgeoning U.S. Geological Survey population at and near the Denver Federal Center.

Colorado Building in Boulder

Office space at moderate rental was found in downtown Boulder's only "skyscraper," the modern Colorado Building at 14th and Walnut Streets. The Society took over the entire top (eighth) floor and a spacious penthouse on a five-year lease. Later as operations expanded, several additional offices were rented on a lower floor. The new quarters were well appointed, required but little alteration to meet the Society's needs, and offered superb views of the close-by Front Range of the Rockies. The Society's new home, considered by all to be a temporary ond pending decisions as to a permanent home, was entirely adequate for the needs at the time. GSA moved in during September 1967 and lived in the Colorado Building for five full years.

The summer of 1967 was a hectic one for the New York headquarters staff. Not only did the accumulated records of nearly 80 years have to be sorted, evaluated, and packed for shipping but also furniture and office equipment too had to be moved, as well as such Society treasures as the massive Penrose library. Contrary to expectations that about half of the existing staff would choose to move to Boulder, only three staff members and the Executive Secretary finally elected to move (and two of the three returned to New York in little more than a year). This meant that equitable separation terms had to be worked out for most of the staff and help given to them in their search for new jobs. It also meant that Ray Becker's first job when he arrived in Boulder, even before the Society's goods had arrived, was to hire, organize, and train an almost entirely new staff. All this was accomplished. The regular business of the Society was also somehow continued. The *Bulletin* fell behind by several months and some other correspondence and mailings were delayed. All preparations for the November Annual Meeting, which strains the staff unmercifully even in normal circumstances, were accomplished on time, partly by the New York staff and partly by the new

one in Boulder. The headquarters operations settled down remarkably fast with the wholehearted team work of the untried but eager and well-chosen staff. By the spring of 1968 the *Bulletin* was back on schedule, and GSA was ready for new challenges in a new setting.

The transition from New York to Denver was a momentous event for the staff if not for the membership. The backup in publication of the *Bulletin* was partly alleviated by having several issues printed in Denver. The Society's long-time regular printer, Lane Press of Burlington, Vermont, continued the printing of other issues. The first issue to be printed in Denver was off the press and ready for distribution when an alert member of the new staff noticed that the cover page bore the notation "Published each month in New York by the Society." New covers had to be printed and applied to 12,000 copies! The error correction was costly but constituted only a tiny fraction of the moving costs and inexpensive in terms of employee pride in helping to establish "their" Society in Colorado.

GSA Builds Its Own Home, 3300 Penrose Place, Boulder, Colorado

In November 1969 the Council authorized search for a permanent headquarters building in the Denver-Boulder area. A new ad hoc Committee on Permanent Headquarters was appointed from among local geologists. It consisted of R. Dana Russell, Chairman; Raymond C. Becker (later replaced by Edwin B. Eckel); B. Warren Beebe; William C. Bradley; John D. Haun; and J. Fred Smith.

Geological Society of America headquarters, 3300 Penrose Place, Boulder, Colorado.

Three years after its authorization and almost exactly at the expiration of the five-year lease on the Colorado Building space, the permanent headquarters building was finished. To disrupt regular business as little as possible, the staff and the Society's belongings were moved in during the weekend of August 26–28, 1972.

The ad hoc committee had begun its work early in 1970. From then until the new home was occupied, the committee members and many of the staff had few dull moments. A professional development consultant, Paul Heffron, was engaged early to provide advice on everything from selection of a site and an architect to almost daily checks on construction progress. The decision to employ such a person proved to be one of the wisest of the many decisions made by this hard-working committee. Again, the choice of a skilled, talented, and understanding architect, Arthur E. Everett, Jr., picked from several candidates by joint decision of the committee, the Society management, and the consultant was a fortunate one.

Numerous available sites of not only raw land but rental properties were examined and carefully considered.

Even the possibility of asking the Society to buy the Colorado Building where it already had its offices and which would have brought in rental income (and manifold problems) was studied in detail. The main requirements considered during the site selection process were (1) reasonable costs of land and construction as compared to future values, (2) a view of the Rocky Mountains to the west, and (3) location on high ground above the flood plain of Boulder Creek and its tributaries. (Much later, after land had been brought and construction started, one of the officers was moved to arrange for an independent study of flood hazards by hydrologists of the U.S. Geological Survey. Fortunately, they found that the site selection group, most of whom were capable map readers themselves, had chosen well.)

The search narrowed rapidly, and a site was selected near the northeast corner of Boulder adjoining the Diagonal Highway between Boulder and Longmont. Approximately two acres of land were optioned and later purchased; requisite soil, topographic, and other studies were made; zoning requirements were met; and at the So-

ciety's request the new access street was named Penrose Place by the City Council. Boulder, which was just beginning a strong antigrowth policy, welcomed GSA because it would employ local residents rather than an imported staff and because its presence would add somewhat to the scientific and cultural atmosphere of a university city.

The architect, the development consultant, the committee, and the staff worked well together from the start. The architect spent many days in study and conference learning much of the character, history, and objectives of the Society. More important, he conferred at length with almost every individual on the staff. From them he learned not only what they were doing but how they interacted with other individuals and groups. With this accumulation of knowledge he designed a building around the needs of the Society and its staff that was not only an architectural triumph but efficient and utilitarian in the extreme. A few very minor alterations have been made since to accommodate changes in operations, but the basic design will continue indefinitely to serve its purposes.

One year after its appointment, the ad hoc committee assisted by Development Consultant Heffron and Architect Everett presented its plans and recommendations, together with a striking scale model of the proposed building, to the Council at its Milwaukee Annual Meeting in November 1970. Following extended discussion of the plans, of the economics of building and owning a Society home, and of the tax advantages of reducing investment income by diverting capital from the stock market to real estate purchase, the Council adopted the committee's recommendations without essential change.

The committee was warmly thanked and nominally discharged, but actually it continued informally for two more years with oversight and advice throughout all phases of construction, furnishing, and decorating of the new home. The Executive Secretary was authorized to purchase the land at the optioned price of about $100,000, to approve plans for a building of about 15,000 square feet with provisions for an added wing at a later date, to approve plans for landscaping and roads, and to approve purchase of interior decoration and furniture. The maximum authorization for all of these purposes was to be $750,000. Of this, $262,000 was to come from the sale of the East 46th Street house in New York and $488,000 from realized capital gains that had accumulated in the Penrose Endowment Fund.

The restriction on total cost was adhered to almost exactly by dint of constant oversight of all phases of design and construction and by sacrifice of some things that seemed desirable but too expensive. Later, the minor alterations plus purchase of the "house next door," described in a separate section, brought the Society's investment in a home close to $1,000,000.

The praises of the Society's home—its esthetics, its architecture, its dignity, and its efficiency as a work place—have been adequately sung in the past and require no further discussion here. Some idea of what it is like can be gained by the accompanying sketches, but a visit to the building itself is required for a genuine appreciation of what it has to offer. A booklet entitled *A Guide for Visitors* is given to all touring the building, and others who are interested in their Society's home may request copies of this booklet from the Executive Director.

The building was formally dedicated at an all-day open house on October 20, 1972. At the dedication, attended by many GSA officers and other Members as well as by about 1,200 other friends, President Luna B. Leopold said:

Perhaps we could have fared well enough in a plain and traditional building. Instead, all concerned wanted something as beautiful as we could obtain, unusual in design and in outlook, something that we hoped would constantly remind all viewers of the grace and symmetry exhibited in earth materials. The many persons associated with the effort were highly concerned with the view, the landscaping, and the ease and comfort of the staff. Because geologists as a group love the work they do, it was hoped that the staff maintained to serve geologists and geology would find, similarly, a pleasant, practical, yet beautiful milieu in which to carry out their daily tasks. (From brochure issued on the occasion of the dedication of the new headquarters building of the Geological Society of America, Boulder, Colorado, October 20, 1972.)

The House Next Door

In November 1972 the Council authorized the purchase of what came to be known as "the house next door." This is a rather attractive brick two-family dwelling only a few feet west of GSA's headquarters land and surrounded by nearly one acre of land. The deed to GSA's property included a covenant prohibiting any building more than two stories high to the west of its building, but there was no protection against undesirable neighbors or use of the adjoining land. To provide such control and to protect its investment in the headquarters building, the Council agreed to buy the house and lot for the surprisingly low price of $90,000. Though originally bought for protection only, the value of the property has increased significantly and the house has continued to be used effectively for various Society purposes ever since its purchase. It appeared to be a good investment in 1972 and should continue to be so indefinitely, although its ultimate uses cannot be predicted.

It is no secret that there have been grumblings and complaints among some Members that the Society spent too much in providing palatial, beautifully landscaped quarters for its employees. Conversely, it is also no secret among the staff that no GSA officer or Member who has ever actually visited headquarters has failed to be favorably impressed. Almost without exception, GSA visitors have voluntarily expressed their pride in being represented by such a building and their satisfaction with the work that is done there on their behalf. Members of the nongeologic public are equally impressed by the award-winning architectural excellence of both building and grounds.

True, GSA's land and buildings cost about $1,000,000 in 1972, about one-tenth of the Society's total assets and almost half of its normal annual budget. Anyone with even a slight acquaintance with land and building values will know, however, that the 1972 investment in a home was sound. No appraisal has been made since the building was occupied, but it is common knowledge that land and construction costs have since skyrocketed through the Denver metropolitan area and more so in Boulder than elsewhere. Office space rentals, always a possible alternative to ownership, have kept pace with rising real estate values in the Denver and Boulder markets. The Society's home was built to endure, both physically and architecturally. It can only continue to increase in value with time.

Part III

Activities, Accomplishments, and Objectives

The Society devotes all of its efforts and resources to a variety of activities, all of them aimed at one objective— advancement of the science of geology. It aids its friends and is aided by them.

The Society's Publications and Its Editors • Meetings • Honors and Awards • Penrose Conferences • Research Grants • Libraries and Depositories • GSA: Good Friend and Neighbor

10

The Society's Publications and Its Editors[6]

Introduction

Need for a publication outlet for papers on American geology was one of the prime reasons for formation of the Geological Society of America in 1888. This was, and still is, stated clearly among the Society objectives—"the promotion of the science of Geology by *the issuance of scholarly publications*, the holding of meetings, the provision of assistance to research and other appropriate means."

Ever since 1889 when the *Bulletin* first appeared the Society has been a leading publisher in the earth sciences. More of its resources, human and material, have consistently been devoted to this activity than to all others combined. Also, more of its prestige is attributable to its publications than to any of its other activities.

In addition to its main journal, the *Bulletin*, which has been published uninterruptedly, the Society launched a second journal, *Geology*, in 1973. Since 1933, it has also published hundreds of books, maps, and charts, some in series, some not, some ephemeral, some permanent additions to knowledge. In total, the mass of the Society's publications is impressive and constitutes a significant part of the geological world's fund of basic knowledge. The various categories and their histories are briefly described below.

In the early years, election to fellowship was generally considered to carry with it the right to publish in the Socie-

ty's *Bulletin*. Consequently, no matter the stringency of technical review, virtually all manuscripts submitted to the Editor were eventually accepted, some of them after extensive condensation. Neither this custom nor the obvious corollary that publication "rights" were restricted to Fellows of GSA was ever formally stated, but they were largely in effect until the early 1930s. When the publication program expanded as a result of the Penrose bequest, both customs were gradually abandoned and manuscripts tended to be judged on their merits rather than on their authors' membership status.

Though some of the earliest papers published in the *Bulletin* discussed such remote subjects as the planetesimal hypothesis, most of the material in GSA publications has had to do with the geology of the Americas, chiefly of the United States and Canada. This restriction was indeed a required one for many years, for the stated purpose of the Society was to promote the science of geology in North America. With expansion of geology's horizons this parochialism was gradually modified; today the Society welcomes papers from all authors in all parts of the world, concerned with all branches of geology.

Those who would complain that GSA publishes too few papers or books within their own personal interests and too much in other fields must remember two things. One, the authors themselves determine the subject matter in GSA publications; and two, so long as the Society strives to

---

6. The year 1980 marked a turning point in many facets of the Society's publication program. For a summary of the more significant changes, see the final section in this chapter.

cover the entire science of geology, it must inevitably print many papers that are of great interest to only a few specialized readers.

With rare exceptions, all material offered to the GSA Editor for possible publication is volunteered. On occasion, an Editor, usually one new on the job, has sought out a paper on a specific topic. As often as not the results have been disappointing and embarrassing, for what can the Editor do with an inferior paper that he has solicited and, at least by implication, committed himself to accept? By skillful use of the review process the Editor can control to some extent the breadth of subject matter in the papers he accepts or can gently steer some manuscripts to other, more specialized journals where they are likely to find more receptive audiences. By and large, however, the Editor must depend on authors to determine the subject matter that ultimately appears in GSA publications.

Throughout its life, the Society has striven to represent all facets of geology, not only in its publications but also in all its other activities. This was easy in the old days when American geologists were few as compared to the number of unsolved or unrecognized geologic problems. Almost every geologic map or report, therefore, presented new facts or theories that were of interest to almost every geologist.

In recent years, however, geologists are far more numerous than in the early years of the Society, and most of them are highly specialized in their interest and professional work, if not in their basic training. Such specialization, by no means unique to geology, poses grave problems to present and future Society Editors. Knowing that no one article in the *Bulletin* or *Geology,* for instance, is likely to be studied by more than a handful of the total readership, how much and what kind of material should be accepted by the Society and preserved in print for posterity?

Until recently, when author-prepared typescript began to be admitted as camera-ready copy for abstracts, microfiche, and other media, virtually all of the Society's publications have been carefully (some authors and others have thought far too carefully!) edited, proofread, and prepared for the printer by well-trained staff editors. Except for informal in-house use and for general instruction on preparation of manuscripts for the main journals, there has never been a GSA style manual such as most publishing houses provide. Instead, the staff depends on a few standard books on writing as authorities, modified as needed for the Society's purposes and as the English language changes with time. GSA "style," however, has remained remarkably constant through the years; most language critics would rate it as good to excellent scientific writing.

In earlier years, GSA provided financial aid to some authors to pay for professional manuscript typing and preparation of illustrations, but this practice continues only for the *Treatise on Invertebrate Paleontology,* a special

case. Again with the exception of the *Treatise,* the staff has never included professional illustrators. This was in the belief that to provide drafting services at all would lead all authors to expect them, thus to greatly increase publication expense. The result of this policy has been that the quality of illustrations published by the Society varies between far wider limits than does the quality of the text.

Pricing and Distribution

Pricing and distribution of its publications to members and others have posed recurrent problems ever since 1933 when the program expanded enormously. Prior to that time, the only publication was the quarterly *Bulletin* which was more than supported by membership dues and modest subscription rates to the comparatively few other users. Ostensibly, the Society prices publications to recover all costs of editing, printing, and distribution, plus in recent years sizeable markups for overhead.

Though the phrase "priced to recover costs" sounds businesslike and convincing, it has been in truth a fiction through most of the Society's publication history. Not until 1958, for instance, were in-house editorial costs added to the manufacturing costs. Arbitrary markups to cover overhead costs were charged from time to time, but these varied widely in percentages applied to equally variable bases. Probably the most important, yet most elusive, factor of all in the pricing of publications was the quantity printed and the estimated proportion of a total edition that would be sold. The greater the number of books printed at one time, the lower the unit cost, but unsold books bring in no income and storage costs money. If the price is set on the expectation that the entire edition will be sold, the Society is more likely than not to lose money in the long run.

Modern, sophisticated methods and machines for cost accounting adopted during the past decade or so have gradually produced much more accurate cost records for the publications program. Taking the publication history as a whole, the Society has certainly operated at a loss in dollars. This is not to say that heavy financial support of its publications has been bad. On the contrary, most geologists would agree that such contributions have been among the most rewarding uses to which the Society's resources have been put.

In reality, some publications are so expensive to produce and appeal to such limited audiences that losses on them are inevitable. Only because of its endowment and nonprofit tax status can the Society accept such works, none of which commercial publishers could afford to publish. Moreover, it provides one of the few outlets that are available to nongovernmental geologists for multicolor geologic maps and other expensive products. Thus, the dual contribution of the Geological Society of America as a publisher is to produce high quality scientific publications at relatively modest prices and to ensure the publication of many basic facts that would otherwise be lost to science.

Sales Efforts

Many efforts have been made through the years to increase sales of the Society's publications to others than members. In 1963, for example, an expert on publication marketing, Henry L. Woodhuysen, was retained as a consultant. He made an earnest effort to seek additional subscribers to the *Bulletin* and to arrange for sales of book publications through dealers, mainly in other countries. He met with only moderate success, and increased sales income failed to exceed the cost of his services plus that of additional staff needed. The consulting arrangement was ended in November 1964. In all fairness, it must be stated that most of the blame for failure should be laid not on the consultant and his ideas but to poor followup by the GSA staff and by foreign agents in terms of deliveries and of billings. Since 1965, as before 1963, the Society has depended on in-house sales promotion, chiefly by means of a widely distributed *Catalog of Works in Print*, printed annually or occasionally as needs and finances warrant. New book publications are also advertised regularly and attractively in *Geology* and are displayed in manned exhibits at most GSA and a few other scientific meetings.

Experience shows that no promotional campaign, inspired or imaginative as it may be, is likely to have much effect on sales of publications like those produced by GSA. The potential market probably varies more or less directly with the world geologic population. Most individuals who want to receive Society publications regularly, however, find it advantageous to join GSA and to get them at member prices. Other potential buyers, such as librarians, either know about Society publications or receive requests for them from their geologist clients and will purchase those they need, with or without advertising.

Publications

Proceedings and Other Society Business Records

Official records of the Society's business and other transactions have been published regularly since its founding. Some of the information was published as required by corporate law, some merely to keep the members abreast of recent developments in their Society. As shown in Appendix C, they have appeared in several media and range widely in content and amount of detail presented, so that the official published record is somewhat spotty and makes for difficult retrieval. Far more complete records exist at headquarters, of course, in the form of Council minutes and their supporting documents, but these are voluminous and not easily available to the casual researcher.

All of the Society's business proceedings—annual reports of officers, finances, elections, record of meetings, and the like—appeared in the *Bulletin* until 1932. As one of the new Penrose-financed series of publications, the *Proceedings Volume of the Geological Society of America* was authorized by the Council in 1933. Except for the years 1961 and 1962, when Society management was in transition and proceedings material appeared in the *Bulletin,* the *Proceedings* volumes were published annually. During much of the 1933–1968 period, these annual volumes were supplemented by informal newsletters, *Interim Proceedings,* sent at irregular intervals to the membership but not published elsewhere. *Proceedings* volumes were discontinued as part of an economy move in 1968. Subsequently, material formerly contained in the *Proceedings* was scattered through several outlets.

Memorials went to the new series of *Memorial* volumes. Citations and responses for the Society's medals and other awards re-entered the *Bulletin,* and reports on annual and sectional meetings went to the new *Abstracts with Programs* series. Annual reports of the officers, including such things as financial statements and lists of investment holdings have gone primarily to the membership, first in a monthly newsletter, *The Geologist,* then in a *GSA News & Information* insert in the monthly magazine *Geology,* where the various parts of the annual report were printed an item at a time in successive issues. Beginning in 1978, the news section was removed from *Geology* and thenceforth sent separately to the membership as *GSA News & Information* on a monthly basis, but still with annual report material scattered through several issues.

In addition to the reports of officers and other annual business material that appears in *GSA News & Information,* this publication regularly reports on all significant Council actions at its semiannual meetings, but by 1979 these lists of actions had become so abbreviated and even cryptic that only a dedicated and perceptive reader would be able to translate them into real knowledge of Society developments.

The scattering of the Society's official reports on its activities in various media, coupled with gross variations in quantity, if not quality, of detail recorded, presents difficulties for any researcher interested in any facet of GSA history. More important, perhaps, the relegation of many published records to ephemeral newsletters spells even greater difficulties for future historians. Material published in regular series—*Bulletin, Proceedings,* and others—is preserved in many public and private libraries, but that in newsletters, whether called *Interim Proceedings, GSA News & Information,* or other names, seldom is.

Bulletin

The *Bulletin,* mainstay of the Society's publication

program and a leader in the field of geologic journals, was conceived by the founders in 1888 and born only a year later, when volume 1 appeared. Originally a quarterly with a press run of 750 copies, it became a bimonthly in 1933 with the impetus of Penrose money and a monthly only a year later. It went to a membership of some 600. Volume 89 (1978) with 1,812 pages went to more than 12,000 members and subscribers throughout the world. In addition to semiannual or annual indexes with each volume, decennial indexes have appeared regularly.

Through much of Society history, a few to several hundred free subscriptions were sent to libraries and other institutions. Though known as exchanges, as described in Chapter 15, emphasis and motive were always on dissemination of geologic knowledge as widely as possible rather than on the Society's receipt of other journals that resulted from some of the "exchanges."

Minor changes in appearance—typographic and editorial style, paper, and methods of typesetting and reproduction—occurred through the years to keep pace with changes in technology and language. Most of the *Bulletin's* physical characteristics, however, survived until 1974 virtually unchanged from the design envisioned and originated by W J McGee, the Society's first Editor. Even today most of McGee's ideas survive, but the *Bulletin* has undergone two major changes in format. The first of these was in 1974, when the page size of volume 85 was enlarged from octavo (about 5¾ x 9 in.) to quarto (8½ x 11 in.). This change brought a flurry of protests from members and subscribers, but the new format was soon accepted in reasonably good grace, particularly because most other scientific journals were also moving toward the larger size. The reason for the change was almost entirely economic, but among several good side effects were provision for more freedom in size and placement of illustrations and tables and easier readability.

The second major change in *Bulletin* format took place in 1979, beginning with volume 90, number 1. (The change, incidentally, was a renaissance of an idea first broached by then–Councilor M. King Hubbert as early as 1948.) The *Bulletin* was henceforth to appear in two parts. Part I, which retained the same page size and appearance as in the past, was to consist of summaries or extended abstracts of technical papers, edited and printed in the same high standards as formerly. Part II was to consist of microfiche reproductions of the entire articles that were summarized in Part I, but photographed from unedited, author-prepared copy. Users could obtain Parts I and II separately or together, or could order hard copy of individual papers from Part II, all at different prices than before.

The decision to change the time-honored method of publishing the *Bulletin* was, of course, based partly on economics—an attempt to cope with the ever-rising costs of editing, paper, printing, and postage that plague all publishers and their readers. More important than economics, however, was the need to find some way to cope with what was predicted to be an ever-increasing flood of manuscripts submitted to the Society for publication. The editorial staff, the officers, and successive Publications Committees had struggled with this problem for years and had tried many methods to reduce the flow, such as stricter selectivity, editor condensations, pleas and threats, all to little avail. In 1968 the National Science Foundation had provided half a million dollars to help relieve the overload with a larger *Bulletin* and editorial staff, but even this proved only a short-term palliative. After several years of thorough and somewhat anguished study, the Council accepted the joint recommendation of the Publications Committee and the staff and voted to change the *Bulletin* to a bipartite format.

Judged from the standpoint of the principal objective—that of coping with too many manuscript submissions—the change proved all too successful! Despite a prolonged and well-designed campaign to prepare both prospective authors and readers for the impending change, the flood of submitted manuscripts slowed abruptly to a trickle. The flow of subscriptions to both Parts I and II from within and without the Society also lessened. Geologists are compulsive writers and must get the results of their research into print, so it was unthinkable that all potential GSA authors had suddenly stopped writing papers. How many authors had simply adopted a wait-and-see attitude toward the new format and how many were finding other outlets were difficult to determine. As told at the end of this chapter, the two-part *Bulletin* was abandoned in 1981 after a two-year trial.

By passive agreement, or explicit Council directives from time to time, the *Bulletin* has never carried editorials. This is but one more bit of evidence that the Society prefers to record the results of research rather than to influence its directions.

Except for notices of its own publications, the Society had never accepted advertisements in the *Bulletin* or other periodicals. The desirability of adding to Society income through paid advertisements had been considered at intervals but had long been avoided for fear of endangering GSA's tax-free status. As one of the many significant changes in publication policy that began to take form in 1980, the Council in May 1981 reversed the long-standing prohibition of advertisements and moved to accept paid advertisements in any of the Society's periodicals (*GSA News & Information*, v. 3, no. 7, p. 105). It also granted permission to Divisions to solicit advertisements for use in their meeting programs.

Content

Scientific Papers. The *Bulletin* is primarily a journal for professional geologists. Since 1889 it had printed thousands of articles and tens of thousands of pages on virtually every facet of geology. Many of the more significant concepts or discoveries have appeared first in the *Bulletin.* Such papers are comparatively rare, however, partly because great new concepts are rare in themselves and also because each new advance in scientific thought is followed by a band-wagon period when scores of workers are moved to do research and to write papers designed to provide detailed evidence either for or against the new concept. Such papers have been submitted to GSA in abundance.

If the subject matter of *Bulletin* papers reflects what geologists are doing and thinking—and writing about—it should follow that the *Bulletin* presents a fine record of advances in geologic thought over the past century. This is true only in a very general way. Not every new advance comes to the *Bulletin* by any means, particularly in an age of specialization and an age in which much new information is proprietary or otherwise withheld from the public. Still, the 90 volumes of the *Bulletin* provide an excellent record of the evolution of general geologic thought and preserve precious details in maps, words, and pictures that are permanently available nowhere else.

Nonscientific Content. Until recently the *Bulletin* always contained a variety of nonscientific matter. From 1889 through 1932, it was the Society's sole publication outlet, hence contained the only permanent printed record of all its business. Later, as the publication program expanded, many of the records were transferred to other outlets, such as the *Proceedings, Abstracts,* and *Memorials* series. Such subjects were moved back and forth from one outlet to another in bewildering and almost whimsical fashion. The various series are briefly described below, and a finding list of material in the chief categories is given in Appendix C.

Not until about 1972 did the *Bulletin* become a pure scientific journal, free of all extraneous business matter.

Geology

Geology, a monthly magazine, was launched by the Society in 1973. It was successful from the start, and even though its costs and consequently its subscription prices multiplied during the next few years, it circulation also increased as it became widely accepted by the profession.

Geology is primarily intended to provide a medium for fast publication of brief but timely articles on any facet of geology, all in an attractive, well-illustrated format. In addition to such articles, it features book reviews, specially solicited by the editor, and a forum section that provides for discussions between geologists on published papers or on other subjects. The magazine also contains some material of more-or-less ephemeral interest that was formerly carried as short notes in the *Bulletin.* Advertisements of Society books and maps are another useful feature.

Geology had its own editor and production staff separate and apart from the staff responsible for producing the *Bulletin* and other Society publications until 1979 when GSA Science Editor Vernon E. Swanson took over the job from Henry Spall, who had been on loan from the U.S. Geological Survey. In 1980 it still had an entirely separate Editorial Board.

The supply of manuscripts submitted to *Geology* has continued to grow ever since its inception, particularly since the *Bulletin* was split into two parts. Quality, interest, timeliness, and diversification of material all remain high. The prognosis for the journal's future as a powerful factor in the advancement of the science is bright.

The Bibliography

The *Bibliography* was one of the first new publication series established by the Society after its inheritance of the Penrose fortune. Except for four of the World War II years, when it was published biennially because of difficulties in obtaining the foreign literature, it appeared regularly each year from 1934 through 1978. It quickly became an indispensable tool for serious investigators and has remained so. Beginning with volume 44 (1979), the Society turned over all responsibility for the series to the American Geological Institute (AGI), which will presumably continue it indefinitely but without financial or other support from GSA.

Through 1968 the Bibliograpy series recorded only non–North American geologic literature and was called *Bibliography and Index of Geology Exclusive of North America;* it was so limited to serve as a companion and supplement to the U.S. Geological Survey's (USGS) intermittent series of bulletins, first called *Geologic Literature of North America,* and since 1919 called *The Bibliography of North American Geology.* Volumes 1 through 30 of the GSA publication were annuals (or biennials for volumes 10 and 11) starting in 1967; volumes 31 through 43 were monthly plus annual cumulative indexes.

Beginning with volume 33 (1969), the series changed in character to worldwide coverage, and its name was shortened to *Bibliography and Index of Geology.* This change came about because the USGS had ceased publication of its North American bibliography. Thus the Society, by then in collaboration with AGI, picked up a service for all geologists, U.S. Government employees or not, that had been performed by the USGS for many years. From 1969 on, the USGS helped ease the strain on the Society's publication budget by buying a large number of subscriptions to the *Bibliography* (at the then very high rates) for distribution among Survey employees in its various offices.

For many years the *Bibliography* had wide distribution by means of low-price subscriptions and free bound copies to the membership. Increasing costs forced gradual tightening of the distribution list, first by requiring Members and soon Fellows to pay for their copies, and later by increasing prices. Today the recorded literature is so voluminous and the subscription price so high that only geologic libraries of institutions and a few companies can afford it.

The first few volumes of the *Bibliography* were compiled and edited primarily by John M. Nickles, who had produced the USGS *Bibliography* for many years. On his retirement from the Survey in 1932, he was employed by the Society at a small salary. Despite age and declining health, he was responsible for producing the first 10 volumes of the GSA *Bibliography*. He died in December 1945. Volumes II (for 1945–1946) through 13 were compiled by his two assistants of pre–World War II days, Marie Siegrist and Eleanor Tatge. Miss Siegrist, who had been hired as a typist in 1935 after several years of clerical work on the XVI International Geological Congress staff, was the driving force behind the *Bibliography* from then until her own retirement in 1973. She and her staff did nearly all of their work at desk space provided in the library of the USGS in Washington.

During the next few years several changes in staff occurred, culminating in a staff composed of Marie Siegrist, Mary C. Grier, and Marcia Lakeman, who were responsible for volumes 15 through 30. These staff members continued on the GSA payroll until their retirements (Grier, 1969; Siegrist, 1973; and Lakeman, 1978), contributing greatly during the transition in character and management of the much enlarged *Bibliography*. (For a more complete statement of Miss Siegrist's life with the *Bibliography*, see *The Geologist*, 1973, v. 8, no. 4, p. 1.).

Collaboration with the Society of Economic Geology. In 1946 an agreement between GSA's Secretary Aldrich and Dr. Alan Bateman, long-time Editor of *Economic Geology*, gave GSA complete responsibility for compilation of the *Annotated Bibliography of Economic Geology*. This semi-annual specialized bibliography, conceived by Waldemar Lindgren, had started publication in 1929. First sponsored by the National Research Council and after 1936 by the Society of Economic Geologists (SEG) with large financial assistance from GSA and other sources, it ultimately consisted of 38 volumes. The last volume, covering entries for 1965, was published in 1967.

Miss Siegrist and her little staff produced much of the material for volumes 19 through 38 of the *Bibliography of Economic Geology*. Particularly in the years immediately after World War II, however, a staff of three bibliographers was unable to cope with the postwar literature explosion plus the additional responsibility to SEG. To solve this problem a large group of part-time abstracters, called "farmees" by Dr. Aldrich, was recruited. This group consisted of a variety of geologists, foreign and American, who

willingly produced hundreds of biblioraphic citations and abstracts for one or both of the two bibliography series. The title pages of all the *Annotated Bibliography of Economic Geology* volumes 19 through 38 listed GSA as the compiler, and individual abstracters were acknowledged in the Introductions. The Economic Geology Publishing Co. was the publisher and distributor.

Collaboration with American Geological Institute. The American Geological Institute (AGI) entered the bibliography picture in 1966 when it established GeoRef. ("Georef" is a coined word meaning approximately "a computer-based geological reference system.") Supported heavily by the National Science Foundation from then through 1974, plus much financial and moral support from GSA, GeoRef provides a system for recording, indexing, and retrieval of references to the geologic literature. GeoRef responsibilities increased enormously when the USGS dropped its bibliographic services in 1969. Though there are many other paying customers for output of items from GeoRef, the chief user of the system was always the GSA *Bibliography and Index of Geology,* which paid AGI for each of the tens of thousands of references that appeared in the *Bibliography* (even those prepared by GSA staff members!). GSA also depends on GeoRef for the annual and decennial indexes of its *Bulletin*.

From its birth GeoRef had more than the usual problems—human, technical, and financial—that beset most new ventures. Both it, and the GSA *Bibliography* consistently lost money and had other difficulties as well. Finally, faced with appropriations problems of its own and seeing no signs that GeoRef would become self-supporting so long as grants were available, the National Science Foundation (NSF) withdrew its support following a 1974 grant.

Prolonged and thoughtful study by AGI, GSA, and USGS, led by George Becraft who happened to occupy strong positions in the publications management of all three organizations concurrently, extended over some months. The ultimate result was a memorandum of agreement between AGI and GSA, signed in mid–1974 by their Executive Secretaries. By this compromise agreement, GSA in effect became a partner and advisor to AGI in the Geo-Ref operation and guaranteed its financial survival until it could become self-supporting. This arrangement was never completely successful or particularly smooth, and it had obvious built-in financial hazards for the Society. It lasted through 1978 when by mutual agreement GSA ceased publication and sponsorship of the *Bibliography* and turned it and its problems over to AGI.

The ending of GSA's support of a powerful research tool for geologists throughout the world was painful. Nevertheless, the continuous publication of the *Bibliography* for almost half a century was one of the brightest chapters in GSA's history of advancement of the science. Moreover, it was one of the activities that would have made R.A.F. Penrose, Jr., proud and pleased with this use of his money.

Memoirs and Special Papers

The Memoir and Special Paper series of book publications, both inaugurated shortly after receipt of the Penrose bequest, compare well with the much older *Bulletin* in adding to the Society's prestige and to the body of geologic knowledge. As part of her duties as Managing Editor, Helen R. Fairbanks set up both book series as well as the *Bibliography* and edited all items in all of them for their first five years.

Since the appearance of both Memoir 1 and Special Paper 1 in 1934, more than 150 Memoirs and nearly 200 Special Papers had been published by the end of 1979. The Memoirs are the more impressive in appearance because they have always been attractively casebound, whereas the Special Papers are characteristically paperbound. In reality the two series differ but little in terms of their subject matter or their impact on the science. As originally conceived, the Memoirs were intended to present thorough, discerning treatments of subjects of great importance and enduring interest. Special Papers, on the other hand, were intended to be shorter, less well-illustrated articles that were too long for the *Bulletin* but on narrower subjects of less permanent value than would be suitable for a Memoir. As time went on, however, distinctions between the two series were gradually eroded. Special Papers, on average, contain fewer pages than do Memoirs, but there are many exceptions on both sides. Some Special Papers are among the most lasting and significant contributions to the literature of geology that the Society has ever produced. Partly in recognition of the growing similarity of the two series, the Special Paper series was abandoned for a brief period during the late 1940s and early 1950s. Special Papers were reinstated within a short time, however, and have continued since.

Treatise on Invertebrate Paleontology

The *Treatise on Invertebrate Paleontology* is one of the more elaborate and prestigious of the Society's undertakings. Moreover, for more than 30 years it has possibly been responsible for more diverse problems—human, administrative, financial, and diplomatic—than any other of the Society's publication series. Founded in 1948 by Raymond C. Moore, with the University of Kansas and the Geological Society of America as co-publishers, it has since developed into a major collaborative enterprise in which more than 200 paleontologists from 18 countries have participated. Twenty-nine volumes, each on a major invertebrate division or zoologically defined unit, such as the

Protozoa, Porifera, or Pelecypoda, had been published (and kept in print) by the end of 1980. Five more parts were still in preparation and several others planned. In addition, such is the ever-accelerating growth in knowledge, several Revisions or Supplements of the original parts had already appeared, six more were in preparation, and needs were seen for many more. Completion date of the project cannot be foretold; if the science of paleontology continues to grow, as it must, so is the *Treatise* likely to continue.

The tangible scientific products of the *Treatise* project—a three–foot shelf of beautifully bound volumes, each with hundreds of meticulously correct line drawings and photographs and with thousands of detailed descriptions of fossil organisms—are available for all to see and to acquire. (The *Treatise* marked another minor first when the Council ruled that members could only acquire their volumes by purchase at the usual discount. This was in contrast to other GSA books that were still sent free to members on request.)

The history of the enterprise, however—the evolution of the series, the human and other problems met and solved, the provisions for the future—has never been compiled and made publicly available. This section is but a thumbnail sketch of a long and complex story. It is condensed from dozens of reports to the Council from the *Treatise* editors, from Publications Committees, and from several special committees. Other sources have included long conservations with two of the three *Treatise* editors, R. C. Moore and Curt Teichert and their staffs, the reading of Moore's last will and testament, and study of numerous other personal or official documents.

Of most used in preparing this summary were a booklet by Moore (1952) and several reports to the 1974 Council by the ad hoc Committee on the *Treatise on Invertebrate Paleontology,* headed first by Alfred G. Fischer and later by J. Thomas Dutro, Jr. Moore's booklet was originally submitted to the Council of the Paleontological Society in November 1951. The 42-page octavo booklet was "distributed" by GSA in 1952. Copies are doubtless still extant in some libraries and personal files, but they are scarce. The booklet is a fine description of the plans and aspirations of the *Treatise* originators.

The idea of updating the old Zittel-Eastman *Textbook on Paleontology* was first broached to the Paleontological Society in 1944 by Benjamin F. Howell of Princeton. That society's Council approved of the proposal (though with some misgivings as to its feasibility) and appointed a committee to explore it further. The committee wisely identified the discovery of the right leader as the paramount requirement. Raymond C. Moore of the University of Kansas was finally invited to assume the task. He accepted and began the planning and organization in early 1948. From then until shortly before his death in 1974, Moore was the driving force behind the entire operation. Without him, or someone much like him if such existed, it is doubtful that the *Treatise* would ever have been born, much less that it would have attained the status it has in the paleontologic world. Autocratic, egotistical, demanding, and brusque as he was to friends and foe alike, Moore was also a great scientist and a man of vision. More important perhaps, he was a good organizer with a knack for getting almost as much work from others as he did from himself.

Moore agreed to accept the *Treatise* leadership only if the Paleontological Society, the Society of Economic Paleontologists and Mineralogists, and the British Palaeontographical Society would agree to act as sponsors. All did so, and each appointed representatives to a Joint Committee on Invertebrate Paleontology with Moore as chairman. Later, the Palaeontological Association, also British, joined as another sponsor. This arrangement continued until 1978 when all four sponsors dropped out and the *Treatise* became a joint project of the University of Kansas and the Geological Society of America. The joint committee provided much helpful support and guidance to the project, particularly during the formative period. In reality, however, the *Treatise* was a one-man (Moore) show from the beginning, and committee members soon learned not to offer advice unless they were specifically asked by the chairman.

Early in the organization phase the Geological Society of America was drawn into the consortium, though not then or later as one of the technical advisors. Specifically, GSA was asked to provide financial support by way of research grants to Moore or to some individual authors for preparation of the innumerable illustrations that would be needed. By some obscure reasoning, GSA's agreement in November 1948 to provide an initial research grant of $25,000 carried with it the responsibility for publishing and distributing the project's products. Thus began the Society's enduring involvement with the *Treatise*. Since the first volume appeared, the University of Kansas and the Geological Society of America have been listed as co-publishers; the university delivers finished books to the GSA warehouse, and the Society pays for printing, binding, and distribution, repaying itself through sales income. Strong differences of opinion have risen from time to time as to the equitability of the arrangement and whether the Society has made or lost money on the *Treatise*, but the general scheme has continued unchanged.

Organization and Management. From its inception, the heart of the *Treatise* project has been in Lawrence, Kansas, in facilities provided by the university through the Paleontological Institute of the Department of Geology. Raymond C. Moore, as editor, led the operation from 1948 until he died in 1974. Curt Teichert joined Moore in 1964, originally as editor for *Treatise* supplements and revisions

of earlier published volumes; later, as Moore's health began to fail, Teichert gradually took over general supervision. Richard A. Robison became assistant editor in mid-1974 and editor in June 1975 when Teichert reached retirement age and left the university. All three editors were on the University of Kansas faculty, but the *Treatise* was their prime responsibility.

In Lawrence, the three successive editors have been aided by a small but able and dedicated full-time staff of editors, illustrators, and secretaries. The job of the Lawrence team, in addition to planning and management of the entire operation, has been to turn authors' raw manuscripts into printed and bound volumes. When necessary, the editors have translated entire manuscripts that arrived in other languages (*The Geologist*, v. 9, no. 2, p. 3, 1974).

On November 17, 1974, the Council discharged its ad hoc Committee on the *Treatise* and established a new standing GSA *Treatise* Advisory Committee. This committee consisted of two outstanding paleontologists and GSA's Executive Director. To serve as the chief link between the *Treatise* and the GSA Council, it was empowered to (a) advise the *Treatise* editors on all policy problems, keeping an eye on funds available, (b) advise on special editorial matters such as acceptance or rejection of manuscripts, and (c) work with the GSA Publications Committee on such problems as reprints, promotion, pricing, and sales. Thus, after 30 years of largely informal and often confusing ties between GSA and the *Treatise* management, the Society at last assumed a strong and leading position in that management. Such leadership could probably not have been attained much earlier for Moore wished to retain all authority in his own hands.

Finances. No exact figures are available, or ever will be, as to the total money input from all sources to the *Treatise* project. Despite its proprietary feelings, however, GSA is certainly far down the list of financial supporters. Both individuals and institutions have contributed much to the *Treatise*. Among the individuals, R. C. Moore did the most. His salary was paid by his University, but he put much of himself into the *Treatise*, and he contributed heavily from his own funds for 30 years. He even provided for its support after his death by a generous legacy, described below.

The several hundred authors of *Treatise* articles contributed their time and knowledge with little or no compensation. In the aggregate, however, the amounts contributed by their institutions in the form of salaries, office and laboratory facilities, and other aid must total several million dollars. Of all these, the University of Kansas surely heads the list of institutions, for it has not only paid the salaries of Dr. Moore and his successor editors for more than 30 years but also contributed very heavily in many other ways. The

National Science Foundation (NSF) supplied more than a million dollars during the years 1958–1975.

Aside from the costs of printing and distribution of *Treatise* volumes, which have almost certainly been more than offset by income from sales over the years, the Society's contributions to the *Treatise* have been relatively small. Research grants to the editor and to some of the authors aggregated something like $40,000 during the years 1948–1957. The Society contributed nothing during most of the 1958–1970 period while NSF support was at its height. As NSF grants began to dwindle in 1971, however, the Society gave small grants to the *Treatise* office in Lawrence in lieu of the editing expense that it would have incurred in Boulder had the *Treatise* been treated like any other GSA book publication.

Raymond C. Moore Legacy. In 1965, Dr. Moore took a giant step in providing for the long-term financial support of the *Treatise* and for a permanent tie between the Society and the University of Kansas. He announced that he and his wife Lillian intended to leave their entire fortune, then valued at about $200,000, to the Geological Society of America. The university was to administer the fund and to provide some guidance in its use, but the intention of the Moores was to provide financial support for the *Treatise*; only after that project was completed was their money to be available for support of other Society publications, preferably on paleontology. The Council promptly assured Dr. Moore, both privately and publicly at the annual meeting banquet, that the Society would be happy to accept the bequest, second in size only to the Penrose bequest, and the responsibilities that would go with it, whenever it should become available.

In late 1973 and early 1974, realizing that the *Treatise* was even then on shaky financial grounds, Moore sought to stabilize it by authorizing his friend, financial manager, and future executor to supply up to $40,000 per year for *Treatise* operations. This money was to come out of Moore's own capital and income while he lived and out of his share of the estate after his death. Moore died in April 1974. The executor was willing to abide by his wishes, even though it meant invasion of the capital, which had by then grown to about half a million dollars. After careful study of Moore's probated will, however, the University of Kansas and the Society lawyers found that none of the Moore fortune could be turned over to the Society until after Mrs. Moore's death.

There the matter stands. At some unpredictable future time, Dr. Moore's dream of generous, continuing support for his *Treatise* and for other Society ventures in paleontology will become a reality. Meanwhile, and to help bridge the gap, the Council from late 1974 through 1980 regularly appropriated $20,000 per year toward *Treatise* preparation.

These sums were in addition to the ongoing responsibility for printing and distribution of the series.

Abstracts

With very few exceptions, a summary of every paper that has ever been presented orally or by poster session at a Society meeting has been recorded in print. This includes abstracts of papers presented at the annual meetings of GSA and its associated societies and at the section meetings. For some years all geologic papers and programs presented at the meetings of the American Association for the Advancement of Science were also published. Abstracts are essential tools for meeting committees in arranging their programs and in deciding which papers to schedule for oral presentation. For almost 60 years, abstracts that were considered unsuitable or for which there was no room on the program were yet scheduled as "to be read by title" and they were placed in print along with all others. This practice was abandoned in 1947.

Not all abstracts are well prepared, and many of them are of only transitory value. By continuing to publish them through the years, however, GSA has added enormously to the store of knowledge, for thousands of ideas or observations have been preserved that would otherwise have been lost. In recent years, abstracts have come to be considered as citable, formal publications, listed in bibliographies. In earlier years they were not so considered.

For some years the Society wished to have "first refusal" for publication of manuscripts on subjects that were first announced by abstracts and oral presentations. This insistence died out gradually, particularly during the 1930s and 1940s when manuscript submittals of all kinds began to proliferate. Nowadays, comparatively few abstracts re-emerge as finished papers, and many of those that do are sent by their authors to other outlets than GSA. Innumerable abstracts, therefore, remain as the only permanent records of much sound geologic work.

As shown in Appendix C, abstracts and meeting programs have appeared in several media through the years— the *Bulletin*, Proceedings, and Special Papers. No convincing rational explanations of the shifts from one medium to another are apparent, but most of them were made for economic reasons as perceived at the time. Since 1969, abstracts have had their own medium. Called *Abstracts with Programs,* this series consists of seven numbers per year, one for the annual meeting, and one for each of the sectional meetings. These are distributed before the meetings (as were most abstracts and programs in former years) to enable attendees to make plans to hear those of their choice.

Memorials

Since the earliest days, the Society has published tributes to its deceased members, commonly including their photographs. Such tributes, called "Memorials," have customarily been actively sought only for Fellows and Correspondents/Honorary Fellows, though some memorials to deceased members are volunteered at times and are published. Not all deceased Fellows have been memorialized; some families do not wish them, and some memorialists, usually professional friends of the deceased, have never completed their tasks. The writing of letters of condolence to survivors, coupled with requests for their wishes as to memorialists, is one of the saddest responsibilities of the Society Secretary, perhaps surpassed only by the recurrent duty of reading the names of deceased friends at each annual business meeting. Well into the 1950s, another of the Secretary's duties was to send telegrams of condolence as well as floral tributes to the families of deceased Fellows, and even to arrange for GSA representation at funerals when possible.

Prior to 1973, memorials were published in the *Bulletin* or in the Proceedings series (see Appendix C); reprints of individual memorials were sent to the families and to the authors. Beginning in 1973, a new series called "Memorials—The Geological Society of America" was initiated. The paperback volumes are published annually or at somewhat less frequent intervals. Editions are limited and sales, mostly to libraries and a few historians of geology, are relatively small. Families of the deceased are given a supply of preprints, attractively printed on good-quality paper. These are produced as soon as each memorial is ready, so they are commonly available to families and to GSA members who request them long before the quasi-annual compilations are published.

The question of whether or not the Society should continue to publish memorials, and if so in what form and about whom, has recurred intermittently for many years. Expectably, the question has tended to rise during periods of economic stress or when submitted technical papers were at flood stages. Of all the studies that were made, that of the Special Committee on Memorials presented to the spring Council meeting in 1951 is one of the more thoughtful and decisive (Council Minutes for 1951, p. 79–93). Its report, by John K. Ambrose, Philip B. King, A. Scott Warthin, Jr., and Glenn L. Jepson, Chairman, is summarized here, as it seems to represent what continues to be the philosophy of devoting Society funds to publication of memorials.
Purposes of memorials:

1. To honor the deceased.
2. To evaluate the geologist and his or her work, and to assemble the testimony (mainly by bibliographies) of eminence in a professional career.
3. To record and preserve facts and contemporary evaluations as source materials for history.
4. To stimulate geologists, particularly the younger ones, by precept and example and thus to increase the strength, dignity, and stature of geology.

5. To "reveal the human aspects and the emotional currents as well as other nonfactual and noneconomic elements in geology as a way of life."

From these and other considerations, the committee concluded that memorials have high value to the Society and beyond as unique scientific and social records of geology not assembled or published elsewhere.

The committee also considered various suggestions that had been made for changing the method of handling memorials. These included (1) use of smaller photographs, (2) shorter biographies, (3) inclusion only of elected officers and medalists, and (4) determination of length according to stature of the geologist memorialized. Each of these alternatives, among others, was thoroughly and factually considered, and each was rejected. The committee concluded that the then current practices in publication of memorials of Fellows and Correspondents were better than any proposed alternatives and should be continued. It added a suggestion that brief obituary notices, without portraits or bibliographies, be published for deceased members. The Council agreed with the recommendations, as would several future Councils agree with similar ones.

Membership Directory

The *Membership Directory,* formerly called "Yearbook," published annually since 1968, does much to relieve the shortage of information on Society activities that appears in other media. Primarily a list of members, with addresses, status, and year of joining GSA, the Directory also contains much other nongeologic information. This includes reproduction of the Constitution and Bylaws; lists of officers, councilors, committees, and past presidents; and meetings schedules, membership statistics, and makeups of the divisions and sections. The *Directory* is intended for purchase by GSA members only, but libraries subscribing to the *Bulletin* may also purchase it.

Miscellaneous and Occasional Publications

In addition to the formal series of publications already described, the Society has also published many other items in many categories. These include Case Histories in Engineering Geology, Reviews in Engineering Geology, stratigraphic correlation charts (which were reprints from *Bulletin* articles), geologic maps and physiographic diagrams, the *Rock Color Chart* (designed for field use), and a few separate unnumbered volumes. For some years, guidebooks for field trips held in conjunction with the annual meetings were also produced for use on the trips and for later sale. This practice was abandoned in 1960 when the Council ruled that the holding of field trips and preparation of guidebooks for them was at the option and responsibility of local committees. As a result, records of subsequent field trips have appeared in several forms and have been published locally by various institutions other than GSA.

In 1953 the American Geological Institute (AGI), then being supported by the National Research Council, GSA, and other societies, started publication of *Geological Abstracts,* a small monthly journal. A forebear of AGI's Geo-Ref information and retrieval system, it depended on volunteers for most of its abstracts and on subscriptions and AGI's sponsors for its money. Shortly after its launching, AGI management proposed that GSA take over the management on a cost plus 10% basis. The GSA Council agreed, as a part of its ongoing support to AGI. Title to the *Abstracts* journal remained with AGI, but the masthead of volume 2 and succeeding volumes listed GSA as "Publisher, on behalf of the (AGI) affiliated societies." The journal was never an outstanding success, and in 1958, after publication of five volumes, it gave way to *Geoscience Abstracts,* which produced eight volumes during the period 1959–1966. This journal was published by AGI with support from NIF; GSA and the other affiliated societies were not involved.

None of the Society's miscellaneous or occasional publications are further described here. In the aggregate, they have added materially to the mass of geologic knowledge recorded in the more formal series, and all can be located in bibliographies or in GSA's publication catalogs (though the latter commonly lists only those publications that are still in print).

The Editorship

Throughout the Society's history, its Editors (called Science Editors or Editors-in-Chief at times) have had nearly as much to do with its welfare and its impact on the science as have its Secretaries. In addition to their influence as editors and guides for the publication program, several Editors have helped directly in the Society's administration by acting as Assistants to the Secretaries or carrying the dual jobs of Secretary and Editor, sometimes for long periods.

From 1897 through 1932, Editors were elected from among the Fellows to serve as officers and to hold seats on the Council as parts of their duties. This arrangement enhanced their influence over the Society's welfare. H. R. Aldrich was never elected as Editor, but by reason of his job as Secretary, he retained his elective place on the Council (and Executive Committee) from 1941 until his retirement. Even after the editorship became appointive rather than elective, the Editors have retained much of their influence on Council deliberations; they sit in regularly on all

THE SOCIETY'S EDITORS

Name	Dates	Title
W J McGee	1889–1892	Editor
Joseph Stanley-Brown	1893–1931	Editor
Charles P. Berkey	1932–1933	Acting Editor
Henry R. Aldrich	1934–1960	Editor-in-Chief[1]
Agnes Creagh	1960–1964	Editor[2]
Herbert E. Hawkes	1965–1966	Director of Publications
Raymond C. Becker	1967	Acting Editor
Edwin B. Eckel	1968–mid-1971	Editor[3]
Bennie M. Troxel	mid-1971–mid-1975	Editor
S. Warren Hobbs	1975–1976	Interim Science Editor
Paul Averitt	1977	Interim Science Editor
Henry Spall	1973–1977	Editor, *Geology*[4]
Vernon E. Swanson	1978–1980	Science Editor[5]

[1]As Assistant Secretary, 1934–1940, and Secretary, 1941–1960, it was understood that Aldrich would be in charge of publications as part of his secretarial responsibilities. The title of Editor-in-Chief came into use gradually and informally.

[2]Fred A. Donath and William C. Krumbein generously assisted Miss Creagh with review of manuscripts while she was carrying the dual responsibility of Executive Director and Editor.

[3]Eckel became Acting Executive Secretary in November 1969 but continued as Editor through mid-1971. He also was Acting Editor for half of 1972, when Troxel left the Society temporarily. Bruce F. Curtis and R. Dana Russel kindly assisted him with review and acceptance of manuscripts for the long periods that he had dual responsibilities.

[4]Peter T. Moyer kindly assisted Spall with the review of manuscripts for long periods.

[5]In March 1981, the Council voted to abandon the salaried office of Science Editor and to establish a group of four volunteer Science Editors, one each for *Bulletin, Geology,* book publications, and the maps and charts series.

Council meetings, and though they have no vote, they, along with the Chairmen of the Publications Committee, have strong advisory voices in all matters having to do with publications.

The Editor's prime responsibility for content, format, and literary excellence of all Society publications has always been paramount among his duties. Every Editor has been aided in technical review of proffered manuscripts by specialists among the membership. Called censors (as they were for many years), peer reviews, members of the Board of Editors, or more recently, Associate Editors, these volunteers have aided successive Editors in their decisions; but the Editor himself has quite rightly always had final responsibility for accepting or rejecting manuscripts. On paper, the records indicate that the Council itself made these decisions well into the 1930s, mulling over each manuscript and formally voting on each. In truth, the Councils were guided by advice from the Editor, who had already reviewed the papers, or had had them reviewed by censors.

Beyond bearing prime responsibility for all publications, the editorial job has been what each Editor made it. Until the Penrose bequest brought financial ease, and with it an enormously expanded publication program, the Editor did everything himself, from accepting papers through making editorial corrections to seeing them through the press. The Editors had some clerical help, of course, but this was generally limited to part-time assistance of a single clerk, with an annual allowance of $250 for salaries and all office expenses! After the Penrose fortune became available and the load increased, the Editor's office gradually enlarged, with more staff and more specialization among them.

W J McGee, 1889–1892, then the chief assistant to Director John Wesley Powell of the U.S. Geological Survey, was one of the founders and the first Editor of the Geological Society of America. He edited only the first three volumes of the *Bulletin* before he resigned the position at the end of 1892. He left the editorship only because he was preparing to leave geology and to follow his leader to start a new career in archeology in the Bureau of American Ethnology, which Major Powell was then founding. (For his memorial, see Knowlton, 1913.)

More than any other Editor, McGee left a lasting imprint, not only on the *Bulletin* but on all future Society publications. His successor, Joseph Stanley-Brown (1914, p. 24) pays tribute to McGee's pioneer work in these poetic terms: "As Minerva sprang, full grown and completely armored, from the brain of Jupiter, so was the Society's publication created—a model of form and efficiency by the sole effort and peculiar genius of one of the most competent and brilliant scientific minds the West has produced. In its essential features, the *Bulletin* stands today as it did at the time of its creation."

Both McGee and Stanley-Brown realized, of course,

that the scientific quality of the *Bulletin* depended entirely on its authors (as aided by the peer-review system); they both gave ample credit to this fact. But it was McGee who designed this journal, prescribing all the details of format, typography, illustrations, paper quality, and so on, not only for the Council and the authors but also for the printer. He sought out and contracted with the firm of Judd and Detweiler of Washington, D.C., to produce the *Bulletin*, thus forming an alliance that was to serve the Society for more than 60 years. His *Publication Rules of the Geological Society of America* was formally adopted by the Council April 21, 1891. These rules, perhaps most easily accessible in one of several reprintings in the *Bulletin*, such as v. 42, p. 266–270, were revised in detail many times through the years. Aside from changes in paper stock and in illustrations that were necessitated by new printing technologies, however, the style and appearance of the journal as laid down by McGee in 1890 remained almost constant until 1974. In that year, a change to larger page size and more modern style was forced by economic and aesthetic considerations. An even more drastic change took place in 1979. The content of the *Bulletin* changed many times through the years, with various combinations of technical papers, proceedings, memorials, annual meeting programs, and the like, but the journal retained the appearance and scientific excellence almost exactly as envisioned by W J McGee at the beginning.

Joseph Stanley-Brown was the Society's Editor for 39 years 1893–1931, far longer than any other. Though relatively unheralded as a geologist or in the annals of the Society, he served both fields for more than half a century. He relinquished the editorship at the end of 1931 only because of his increasing burdens as Chairman of the Finance Committee, a group that had been established to cope with the realities of the Penrose bequest. He retained the chairmanship until his death in 1941. With a small committee of colleagues who knew more about investments than most GSA members, Stanley-Brown took the lead in establishing a sound investment policy and most of the other changes that had to be made in response to the sudden need for wise management and use of a large fortune.

For all his 48 years of devoted service as Editor and chief financial advisor, Stanley-Brown received small honoraria but no pay and peculiarly little acclaim. No memorial to him was ever prepared, but at the 1941 annual meeting, shortly after his death, a suitably inscribed gold watch was presented posthumously to his family, together with heartfelt laudatory comments by President Berkey, Treasurer Mathews, and Assistant Treasurer Beatrice Carr. (*Proceedings for 1941*, p. 75–78.) The watch had been intended for presentation in person to mark the Society's indebtedness for a lifetime of service, but death had intervened only a month before the meeting.

The best summary of Stanley-Brown's life and of his

Editors

W J McGee, 1853–1912; Editor, 1889–1892.

Joseph Stanley-Brown, 1859–1941; Editor, 1893–1931.

Herbert E. Hawkes, 1912–   ; Director of Publications, 1965–1966.

Bennie M. Troxel, 1920–   ; Science Editor, 1971–1975.

S. Warren Hobbs, 1911–   ; Interim Science Editor, 1975–1976.

Paul Averitt, 1908–   ; Interim Science Editor, 1977.

Henry R. Spall, 1938–   ; Editor, *Geology*, 1973–1977.

Vernon E. Swanson, 1922–   ; Science Editor, 1978–1980.

connections with geology is that by Fairchild (1932, p. 172–173); the best evidence of his own editorial philosophy and of his ability as a writer is by Stanley-Brown himself (1914). Educated at Yale and later at Heidelberg, his first geologic job was as secretary to the great explorer, Major J. W. Powell in 1876–1879. A little later, Powell, then Director of the infant U.S. Geological Survey, lent him to President Garfield to serve as his personal secretary. After Garfield's assassination, Stanley-Brown was disenchanted with politics and returned to the Survey where he worked intermittently for a few years, then gained expertise in management, political and otherwise, in the Alaskan fur seal industry. From there he became interested in the railroad business through tycoon E. H. Harriman, which led him directly to his real life work in banking and investments. Fairchild noted that, like Penrose, he foresaw the Great Depression, cashed in his holdings at the top of the boom, and retired in comfortable wealth in January 1929. In his own biography (*Who Was Who in America*, v. 1, p. 1171) he does not even mention his work for GSA, but throughout that long and kaleidoscopic career, he devoted most of his leisure time to its welfare, first as a stern and forthright Editor, later as its financial guide, and always as a wise and able Councilor.

Other Editors

Several of those on the list of Editors, or Acting Editors, also served as Secretaries: Berkey, Aldrich, Creagh, Becker, and Eckel. They and their regimes are sufficiently characterized in Chapter 7.

Herbert E. Hawkes, 1965–1966, was appointed as Director of Publications as an experiment in absentee editorship. A professor in geochemical prospecting at the University of California, Berkeley, Hawkes agreed to devote half time to GSA publications. He was to visit New York headquarters intermittently to keep some personal touch but was to do most of his editing work in Berkeley. Hawkes managed to keep manuscripts and sound judgments on them flowing but the flow was irregular, particularly during a long period when he was teaching in England. In addition to assessing manuscripts, Hawkes found time to do some significant comparative studies of the whole problem of geologic publications and their use in research.

On the whole, the experiment was disappointing. There were too many other demands on the Editor, the movement of manuscripts was necessarily awkward and he could not possibly take a meaningful part in the day-to-day business of turning raw manuscripts into finished publications, all of which had to be done in New York. The experiment in absentee editorship was abandoned toward the end of 1966 when Hawkes resigned, and Executive Secretary Becker added the editorship to his other duties. The concept of appointing an absentee Editor surfaced again in 1977 when the publications committee proposed that both the Science Editor and the Editor for *Geology* be part-time appointees. The idea was dropped for the moment, but resurfaced in 1980–1981 (see end of this chapter).

Bennie M. Troxel, 1971–1975, came to the Society after long experience in research, administration, and editing, principally with the California Division of Mines and Geology. His tour of duty was marred by a six–month gap when he returned to his former employer, but he survived the editorship longer than any Editor had since Henry Aldrich. During his regime the magazine *Geology* was launched, and partial success was achieved in coping with an ever-increasing flood of manuscripts by reducing the lengths of those accepted.

The four most recent Editors—S. Warren Hobbs, Paul Averitt, Henry Spall, and Vernon E. Swanson—all came to the Society from the U.S. Geological Survey. Hobbs and Spall, in fact, were outright gifts from that organization, which continued to pay their salaries while they worked for GSA on a half-time basis. This unsung and hitherto unpublicized generosity was but one of many that have characterized GSA-USGS relations since 1889 when W J McGee was similarly "lent" to GSA. Both Hobbs and Averitt, a recent retiree from the Survey who also worked half time but on salary, were Interim Science Editors and were simply holding the fort during a long and often frustrating search for a full-time incumbent. Their tasks were somewhat eased when the Executive Director assumed most of the in-house management duties of the publications program, leaving the Editors with the prime duty of manuscript selection.

A long search for a "permanent" full-time Editor finally culminated in late 1977 with selection of Vernon E. Swanson. He took early retirement from government employment and became Science Editor early in 1978. In his first two years on the job he showed great promise of fulfilling one of the Society's greatest needs, but he became one of the casualties in the 1980 general reorganization of the publications program and left Society employment in November 1980. The post of Science Editor was left vacant pending decisions on major changes in the program and its administration.

Editorial Staff

Since 1932 when the volume of GSA publications began to expand, the actual work of turning words into print has been done by an editorial staff. It is headed by an all-around expert with the general or specific title of Managing Editor, under the general direction of the Science Editor. Helen R. Fairbanks, senior author of the superb Penrose biography (Fairbanks and Berkey, 1952) left her job as Managing Editor in 1938, resigning from the Society. Her position was abolished as it was believed that Editor-in-Chief Henry R. Aldrich could handle both jobs. In practice, Agnes Creagh, who had joined the staff a year earlier as an Editorial Assistant, took over many of the Managing Editor's duties from the start, was formally named to the position in 1944, and continued there throughout the rest of Aldrich's regime. With able assistance from Barbara Flynn, Miss Creagh continued to carry most of the responsibilities of Managing Editor until she resigned in 1964, even though she was also the Science Editor and later Executive Secretary during most of the period. She was succeeded as Managing Editor by Martin Russell in 1964. He left the Society when it moved to Boulder but generously spent enough time in Colorado to help orient the new publications staff. Josephine K. Fogelberg then served with distinction as Managing Editor for 13 years until she retired in 1980. One of the first of the new Boulder staff, she recruited and trained virtually all of the editorial staff during that entire time and set and maintained editorial standards. She conceived and/or activated innumerable innovations in the Society's publications program during her term in office.

The staff has varied in size depending on changes in the publications program. It contains editors, secretaries, proofreaders, typists, photographers, and others, most of them with several skills. The individuals, and even the rest of the Managing Editors are not listed here, but current incumbents are commonly listed in mastheads of the *Bulletin* and *Geology*, as well as in annual reports. Without the high skills and devotion of these editorial staffs through the years, GSA's publications would not have the volume or stature in the scientific world that they do.

By the end of 1980, though the editorial staff structure remained much as sketched above, increasing amounts of the detailed "blue-pencil" editing were done on contract by outside editors.

Technical Reviewers

Through the years the incumbent Editors' lives have been bearable only because they were free to call on the entire membership for help in review and acceptance or rejection of manuscripts. Many hundreds of members, commonly experts on the specific subject matter, have been asked to review individual papers, to offer suggestions to their authors, and to advise the Editor, directly or through Associate Editors as intermediaries, as to their contributions to the science. Nearly all potential reviewers have accepted their assignments gracefully, even though they have perforce interrupted their own research. Other than inner knowledge that they have helped the Society and the science, the reviewers' only reward has been the inclusion of their names, in fine print, in annual reports.

The Society's insistence on peer review of all submitted manuscripts brings pain to some authors, particularly those whose papers have already been reviewed in their own institutions. More than any other factor, however, the system is responsible for the generally high quality of the Society's publications and for thier standing in the literature of geology.

Publication Problems of 1980 and Their Solution

Major problems in the Society's publications operations developed in the early spring of 1980. Collectively, they soon became known to large segments of the Society as the "publications flap." Before they were solved, the in-house editorial staff had been materially reduced by resignations, dismissals, and retirement. Morale of most of those who remained on the rolls was lowered. Far more influential on GSA future directions than staff losses, however, were far-reaching changes in the overall policy and philosophy of the publication program.

Details of the "publications flap"—its character, causes, personalities, and the emotions that shook not only those directly involved but innumerable GSA members—will fade with time and are not recorded here. Because the changes in publication policy are likely to have profound effects on the Society's health for many years into the future, however, it is necessary to summarize them here.

Because the immediate problems involved the interrelations of editorial staff, management, and the officers, the Headquarters Advisory Committee, chaired by John W. Rold, was asked to make a preliminary study. It did so with both wisdom and dispatch. Its report to the Executive Committee and Council in July 1980 tended to distribute blame for the festering publication problems equally between leadership, management, and staff. It recommended a few immediate actions designed to alleviate the situation; most of these were accepted and applied. The Committee's chief recommendation, however, asked for appointment of an ad hoc, blue-ribbon committee, charged with making a thorough study of the entire publications program—purpose, structure, products, and economics.

The "publications flap" simmered through the summer and early fall of 1980. Executive Committee, Council, and headquarters were deluged with telephone calls and letters from the membership and even many geologists outside of GSA as they became aware that something was wrong with the Society's publication program. The worries came to a head at the Atlanta Annual Meeting in November at a special meeting of the Publications Committee, the Associate Editors of both journals, and numerous others. This meeting was followed by unusually large attendance, and extended discussions, at the annual corporate meeting. Neither of these groups could accomplish changes themselves, mainly because a preponderance of mail ballots had already been cast for the routine election of officers and transaction of other Society business. The message from those who took part in the meetings was clearly heard by the Council, however.

One of the more urgent actions of the Council's formal meeting in Atlanta was to effectively kill the two–part *Bulletin* concept which had been in effect for only two years and which was at the root of much of the so-called publications flap. Announcement of the change was made during the Atlanta meeting and published shortly thereafter. As might have been predicted by those who were close to the situation, the promised death of the two-part *Bulletin* resulted in resumption of a normal flow of full–length manuscripts within a few weeks.

Also during its meeting in Atlanta, the Council acted on the Headquarters Advisory Committee's main recommendation and appointed a Special Publications Committee. Headed by Brian J. Skinner, Yale professor and long-time editor of *Economic Geology*, it included geologists who had first-hand knowledge of publications problems not only in GSA but in other societies; it also included some who were deeply involved as constructive critics of the directions that the Society had taken. In addition to Skinner, the Committee consisted of Robert G. Blackadar; Gregory A. Davis; E. R. Ward Neale; Steven S. Oriel; Thomas J. M. Schopf; Arthur A. Socolow; William A. Thomas; Fred A. Dix, AAPG, conferee; and Christopher G. A. Harrison, AGU, conferee (GSA News and Information, v. 3, no. 1, p. 1, 1981).

The Special Committee did a superb and thorough job in a remarkably short time. Its members communicated directly with scores of those who knew parts of the myriad facts needed for rational solutions to knotty problems, and consulted voluminous financial and other records. Even more significant, they sought advice from the entire membership as to what it really wanted in GSA's publications. Nearly half of the members responded, a far higher response than to any othe opinion poll in the Society's history.

The Special Committee's far-ranging and thoughtful report was made to the Council at its May 1981 meeting; it is summarized by the chairman in *GSA News & Information*, 1981, v. 3, no. 7, p. 105. Council immediately accepted virtually all of the recommendations, as reported by President Howard Gould in the same issue.

Principal decisions made by the Council were:

1. The *Bulletin* would return to a one-part, printed, monthly journal with full-length articles. Part II (longer articles published in microfiche) would be abandoned, but the complete one-part *Bulletin* would be available to subscribers in microfiche.

2. The *Bulletin, Geology,* and *GSA News & Information* would to to all dues-paying members and at significantly lower net costs than had obtained earlier.

3. *Abstracts with Programs* and the *Membership Directory* were to be continued, with no changes in price.

4. The office of a single, salaried Science Editor was abolished (for the first time since 1934). Instead, four external Science Editors, one each for the (1) *Bulletin,* (2) *Geology,* (3) books, and (4) the map and chart series were to be appointed. These Science Editors, all to be volunteers but with secretarial and travel expenses paid by the Society, were to have complete responsibility for scientific content of their domains. Production and distribution of all publications were to remain in the hands of a reduced headquarters staff.

5. The Committee on Publications was to be restructured and greatly strengthened. To be chaired by a Councilor, it was to contain the four external Science Editors, plus three members-at-large to be appointed by Council.

These five major changes in policy were backed by several minor ones, designed to relieve some but by no means all of the economic problems that have plagued the publications program for many years. These included (1) acceptance of commercial advertising in periodicals, (2) reinstitution of voluntary page charges for the *Bulletin* and *Geology,* and (3) acceptance of certain bank credit cards for purchase of publications, dues payments, and preregistration at meetings.

Time alone will tell which of the many changes in the GSA publications program will survive and what effect they will have on the health of the Society and of geology. It is to be fervently hoped that the results will be positive.

11

Meetings

Annual Meetings

Next to the publication program (or perhaps ahead of it in some minds) the GSA annual meeting stands at the top of the Society's list of means to promote the science of geology. Freedom to hold meetings exclusively devoted to geology was one of the prime reasons for leaving the American Association for the Advancement of Science. The original Constitution mentioned only the "advancement of geology" as the new Society's purpose, but meetings and publications were clearly in the minds of the founders; both were recognized as stated objectives in later versions of the Constitution.

The first annual meeting, held at Cornell University in December 1888 and attended by but 13 geologists, was devoted entirely to the business of organization; no technical papers were presented. During the following summer when the infant Society met in Toronto in conjunction with the AAAS, a few scientific papers were read (Fairchild, 1932, p. 117; and *Bulletin,* v. 1, p. 15). At the second annual meeting in New York City, December 1889, the custom of emphasizing scientific papers rather than Society business as the chief *raison d'etre* for annual meetings became firmly established. It was never to be displaced.

Scientific Sessions

Two activities, one formal and the other informal, are the principal events at each annual meeting. Together they go far toward advancing the science of geology. The main formal activity is the reading of scientific papers. From the original 6 papers presented at the second annual meeting, the number has grown to 1,000 or more presented at numerous concurrent sessions. Individual papers are grouped as to topic, as are planned invited symposia, for the benefit of special-interest audiences. In recent years, orally presented papers, commonly illustrated with slides, are supplemented by poster sessions, described below. Schedules are crowded, and not all authors are polished speakers or professional illustrators. Discussion time is always limited or nonexistent. Even so, the orally presented papers at GSA, backed by published (and citable) abstracts, represent one of the best ways yet found to spread broadly new knowledge of geology. Opportunity for both older and younger geologists to appear in person before the geologic public is a secondary but highly important by-product of the scientific sessions.

The informal activity of greatest scientific importance is the opportunity that annual meetings provide for individuals to get together in large or small groups, to greet old friends or to make new ones, to discuss current investigations or to try out new theories. This social activity takes place everywhere—in hotel corridors or lobbies or guest rooms, in bars or at cocktail parties (which are many), in exhibit areas, or before and after (or often during) scientific sessions. The chance to take part in the activity is the chief reason for attendance by innumerable geologists and their spouses. The Annual Meeting of the Geological Society of America and its Associated Societies is truly a gathering of the clan.

Complaints about excessive numbers of papers scheduled for annual meetings, and the attendant "3-ring circus" atmosphere created by the need for multiple sessions, are widespread and numerous and have been for many years. Each Joint Technical Program Committee faces the problem afresh and copes with it in its own way; occasionally the Council tries to help by setting arbitrary limits on numbers of papers, of sessions, or of both, yet the problem persists.

The strongest argument against limiting numbers of

papers to be accepted for oral or poster-session presentation is the perennial one that presentation of a scientific paper is considered to be a travel ticket to the GSA meeting. Geology departments, geological surveys, and companies tend to be more free with expense money for geologists appearing on the scientific program than for those who wish to attend for other reasons, and some authors cannot afford to attend a GSA meeting at their own expense. Yet the notion that limitations on numbers of papes accepted would cut overall attendance significantly is patently false. Roughly 13% of those who attend annual meetings present papers. This percentage remained almost constant over the period 1946–1976 (Council Minutes for 1977, p. 19). It has increased somewhat since. For example, for the 4,826 people who attended the Toronto meeting in 1978, a total of 1,193 papers and poster sessions were presented. On the unlikely assumption that all of the speakers have had their ways paid solely to give papers, these figures mean that the great majority of attendees have managed to find their way to meetings for other reasons than paper and poster presentation.

Poster Sessions and Discussion Papers. Poster sessions, as an alternative to oral presentation of scientific papers or to the graveyard category of abstracts "to be read by title," were formally introduced by the Society at the 1974 Annual Meeting in Miami. Beginning with the Atlantic City meeting in 1969 and continuing through 1973 the poster sessions concept was preceded by informal provision for discussion papers at each annual meeting. These were never very successful and attracted only a few authors. Possibly their lack of success was more attributable to poor planning than to other factors, for "discussions" tended to be scheduled on the fringes of large milling crowds of people, few of whom were interested in learning about the finer points of any specific research project. Adverse experience with the discussion papers led to somewhat more rigid structuring for poster sessions. Abstracts of each poster presentation are published in advance in *Abstracts with Programs.* Authors are assigned booths with adequate wall space to exhibit their material— geologic maps, diagrams, data tables, and the like. Exhibits remain on view for at least half a day to permit detailed study and note-taking, and authors are available at stated hours to discuss their findings with anyone interested.

Though billed as a totally new approach to scientific communication and one in which GSA was among the leaders, the poster-session idea was not at all new. In fact, the first recorded seed was sown by William Morris Davis in 1915:

Professor Davis presented a resolution looking to an arrangement of program of the Society, so that there might be an exhibit session, the idea being that time might be allotted to each man who had an exhibit to present his material to those most strictly interested, under conditions that would insure a chance to discuss the matter with him. Motion was carried to the effect that this Council recommend to the new Council the adoption of this suggestion in some suitable form. The Davis resolution, touching special program for announcing exhibits and informal discussions, as handed in, is as follows:

Resolved, that the Council be requested to devote an afternoon at an early meeting to an exhibition or conversazione at which members may set forth on convenient tables the products of their work and explain them informally to such others as may desire to listen, and that exhibitors and exhibits be announced on the program. [Council Minutes for 1915, p. 343.]

From 1915 to 1964, or for nearly 60 years, the seed lay dormant with no mention of it in any written records. Versions of it may have been suggested from time to time in unrecorded sessions of program committees hard pressed to cope with floods of proffered abstracts.

The first recorded suggestions, other than that of W. M. Davis, came from two fellows, I. Gregory Sohn and Herbert E. Hawkes, who in 1964 independently urged the Council to consider adding poster sessions to the program for the 1964 Annual Meeting in Miami.

Regardless of its origin, the concept of poster sessions was an idea whose time had come by 1974 (Maugh, 1974). Initially regarded askance by some authors as unfair relegation to second-class citizenship, many authors soon learned that the advantages far outweighed the disadvantages. Instead of fleeting 15-minute exposures of their faces, ideas, and illustrations to mixed audiences in typical oral presentations, authors had time for personal discussions with those really interested in their work and results. Today, poster sessions are regular and popular parts of all annual meetings of the Society and of some of its sections. They are well characterized by Severson and others (1979), whose paper should be reread by all potential contributors.

Meeting Statistics

Pertinent statistics on all annual meetings are summarized in the accompanying table. From 1888 through 1947, the meetings were held between Christmas and New Year's. This schedule discomfited some members and more families, but it seemed necessary, partly from habit and partly because Christmas week was traditionally an academic holiday period. Finally, a change to November dates was adopted in 1948 amid strong feelings for and against. Perhaps the change marks the time when the stature of geology and of the Society had at last grown to the point where attendance at a scientific meeting could be deemed more important than classroom or other professional work.

Summer Meetings

As part of their efforts to ease separation from the parent association, the founders provided in the Constitu-

GEOLOGICAL SOCIETY OF AMERICA ANNUAL MEETING STATISTICS[a]

Date	Location	Meeting Registration	GSA Members Registered[b]	Papers Presented[c]	Discussion Papers (1969–1973) or Poster Papers (1974–1980)	Total Membership
Dec. 27, 1888	Ithaca, NY	13	13			115
Dec. 26–28, 1889	New York, NY	60		6		173
Dec. 29–31, 1890	Washington, DC	66				197
Dec. 29–31, 1891	Columbus, OH	23				213
Dec. 28–30, 1892	Ottawa, Canada	29				219
Dec. 26–28, 1893	Boston, MA	51				229
Dec. 27–28, 1894	Baltimore, MD	63		27		226
Dec. 26–28, 1895	Philadelphia, PA	57		26		226
Dec. 29–31, 1896	Washington, DC	76		41		233
Dec. 28–30, 1897	Montreal, Canada	31		31		242
Dec. 28–30, 1898	New York, NY	77		43		237
Dec. 27–30, 1899	Washington, DC	70		52		239
Dec. 27–29, 1900	Albany, NY	51		30		248
Dec. 31, 1901 Jan. 1, 2, 1902	Rochester, NY	39		23		245
Dec. 30, 31, 1902 Jan. 1, 1903	Washington, DC	112		80		256
Dec. 30, 31, 1903 Jan. 1, 1904	St. Louis, MO	49		38		253
Dec. 29–31, 1904	Philadelphia, PA	85		61		259
Dec. 27–29, 1905	Ottawa, Canada	39		51		271
Dec. 27–29, 1906	New York, NY	133		86		283
Dec. 30, 31, 1907	Albuquerque, NM	28				294
Dec. 29–31, 1908	Baltimore, MD	157	120	72		294
Dec. 28–31, 1909	Cambridge, MA	97		55		305
Dec. 27–29, 1910	Pittsburg, PA	95		42		319
Dec. 27–30, 1911	Washington, DC	140				320
Dec. 28–31, 1912	New Haven, CT	108		67		342
Dec. 30, 31, 1913 Jan. 1, 1914	Princeton, NJ	266	131	63		380
Dec. 29–31, 1914	Philadelphia, PA	192	117	59		390
Dec. 28–30, 1915	Washington, DC	240	136	80		389
Dec. 27–29, 1916	Albany, NY	193	107			415
Dec. 27–29, 1917	St. Louis, MO	84	43	52		416
Dec. 27–28, 1918	Baltimore, MD	153	119	61		432
Dec. 29–31, 1919	Boston, MA	214	110	72		431
Dec. 28–30, 1920	Chicago, IL	213	120	88		465
Dec. 28–30, 1921	Amherst, MA	225	119	83		457
Dec. 28–30, 1922	Ann Arbor, MI	230	113	77		488
Dec. 27–29, 1923	Washington, DC	358	171	65		505
Dec. 29–31, 1924	Ithaca, NY	214	102	62		514
Dec. 28–30, 1925	New Haven, CT	343	148	65		519
Dec. 27–29, 1926	Madison, WI	363	115	85		534
Dec. 29–31, 1927	Cleveland, OH	369	170	108		558
Dec. 26–29, 1928	New York, NY	722	221			565
Dec. 26–28, 1929	Washington, DC	583	224	85		591
Dec. 29–31, 1930	Toronto, Canada	519	165	94		613
Dec. 29–31, 1931	Tulsa, OK	721	153			645
Dec. 28–30, 1932	Cambridge, MA	512	174			669
Dec. 28–30, 1933	Chicago, IL	571	185			666
Dec. 27–29, 1934	Rochester, NY	524	191			658
Dec. 26–28, 1935	New York, NY	923	236			669
Dec. 29–31, 1936	Cincinnati, OH	600	179			687
Dec. 28–30, 1937	Washington, DC	1,135	292	123		700
Dec. 28–30, 1938	New York, NY	1,200	270	114		734
Dec. 28–30, 1939	Minneapolis, MN	654	151	81		772
Dec. 26–28, 1940	Austin, TX	669	144	100		766
Dec. 29–31, 1941	Boston, MA	626	197	70		776
Dec. 29, 1942	New York, NY	Business meeting held in Society's headquarters during war years.				795
1943	New York, NY	Business meeting held in Society's headquarters during war years.				810
1944	New York, NY	Business meeting held in Society's headquarters during war years.				814
Dec. 27–29, 1945	Pittsburgh, PA	773	254	140		930
Dec. 26–28, 1946	Chicago, IL	919	278	118		1,053
Dec. 29–31, 1947	Ottawa, Canada	909	278	121		1,079

Edwin B. Eckel

GEOLOGICAL SOCIETY OF AMERICA ANNUAL MEETING STATISTICS[a]

Date	Location	Meeting Registration	GSA Members Registered[b]	Papers Presented[c]	Discussion Papers (1969–1973) or Poster Papers (1974–1980)	Total Membership
Nov. 11–13, 1948	New York, NY	1,593	539	150		1,313
Nov. 10–12, 1949	El Paso, TX	1,151	393	141		1,651
Nov. 16–18, 1950	Washington, DC	2,027	748	141		1,932
Nov. 8–10, 1951	Detroit, MI	1,416	601	151		2,290
Nov. 13–15, 1952	Boston, MA	1,689	723	191		2,667
Nov. 9–11, 1953	Toronto, Canada	1,813	695	197		3,020
Nov. 1–3, 1954	Los Angeles, CA	1,431	704	236(?)		3,352
Nov. 7–9, 1955	New Orleands, LA	1,504	826	223		3,948
Oct. 31, Nov. 1–2, 1956	Minneapolis, MN	1,228	659	183		4,274
Nov. 4–6, 1957	Atlantic City, NJ	1,708	913	249		4,593
Nov. 6–8, 1958	St. Louis, MO	1,738	1,189	285		4,869
Nov. 2–4, 1959	Pittsburgh, PA	2,102	1,053	277		5,156
Oct. 31, Nov. 1–2, 1960	Denver, CO	2,702	1,475	348		5,565
Nov. 2–4, 1961	Cincinnati, OH	2,225	1,214	349		5,739(?)
Nov. 12–14, 1962	Houston, TX	2,242	1,797	348		6,082
Nov. 17–20, 1963	New York, NY	2,892	1,962	335		6,412
Nov. 19–21, 1964	Miami Beach, FL	2,045	1,568	400		6,640
Nov. 4–6, 1965	Kansas City, MO	2,371		362		6,913
Nov. 14–16, 1966	San Francisco, CA	3,352		474		7,257
Nov. 20–22, 1967	New Orleans, LA	3,750		429		7,488
Nov. 11–13, 1968	Mexico City, Mexico	2,808		502		7,782
Nov. 10–12, 1969	Atlantic City, NJ	3,356	1,360	493	33	7,823
Nov. 11–13, 1970	Milwaukee, WI	3,430		505	21	8,358
Nov. 1–3, 1971	Washington, DC	4,318		543	11	8,492
Nov. 13–15, 1972	Minneapolis–St. Paul, MN	3,065		505	15	9,429
Nov. 12–14, 1973	Dallas, TX	3,261		647	5	10,319
Nov. 18–20, 1974	Miami Beach, FL	3,616		814	98	12,214
Oct. 20–22, 1975	Salt Lake City, UT	3,733	2,026	713	39	11,954
Nov. 8–10, 1976	Denver, CO	5,351	2,615	852	68	12,191
Nov. 7–9, 1977	Seattle, WA	3,956	1,797	718	138	12,339
Oct. 23–26, 1978	Toronto, Canada	4,826	1,856	1,038	155	12,404
Nov. 5–8, 1979	San Diego, CA	4,574	2,924	823	222	12,166
Nov. 17–20, 1980	Atlanta, GA	4,285	2,758	1,312	244	12,603

[a]Compiled by Annual Meeting Department, GSA headquarters, largely from *Proceedings* volumes and from published programs of individual meetings.

[b]No reliable records of total attendance or of GSA member attendance are available prior to 1913. Total registrations listed for 1889–1912 may actually represent GSA members only. Registration figures for 1913–1931 as given by Fairchild (1932, p. 124–125) differ markedly from those given here and have been disregarded.

[c]Figures for papers presented represent counts of titles or abstracts in meeting programs; some of these may not have been presented orally; volunteered, invited, poster and discussion papers are included, but presidential addresses are not.

tion that GSA should join with AAAS in its annual summer meetings. As shown below, 14 such joint meetings were held from 1889 through 1902. The summer meetings were never very well attended by the geologists, partly because the meetings broke into the field season and partly because GSA held its own annual meetings independently. When AAAS itself abandoned summer meetings with the one of 1902, the GSA Council chose to interpret the Constitution literally and promptly withdrew from the joint meetings arrangement. The Constitution was altered in 1905 to fit the facts—no joint meetings and no summer meetings. One final summer meeting was held ten years later. This was a special meeting in Berkeley and Stanford, California,

in 1915 in conjunction with the Cordilleran Section and the Paleontological and Seismological Societies. (Fairchild, 1932, p. 118.) It was designed to draw the geologists of the western United States and Canada closer into the fold.

Joint meetings with AAAS were as follows: 1889, Toronto, Ontario; 1890, Indianapolis, Indiana; 1891, Washington, D.C.; 1892, Rochester, New York; 1893, Madison, Wisconsin; 1894, Brooklyn, New York; 1895, Springfield, Massachusetts; 1896, Buffalo, New York; 1897, Detroit, Michigan; 1898, Boston, Massachusetts; 1899, Columbus, Ohio; 1900, New York, New York; 1901, Denver, Colorado; 1902, Pittsburgh, Pennsylvania.

As recorded by Fairchild (1932, p. 125), maximum

attendance by GSA geologists at any of these meetings was 53; papers presented ranged from six to nineteen. Most meetings were for a single day.

Meetings Management

The organizing and holding of annual meetings is a big and complex business. Nowadays the budget for each one is more than $100,000. (The total of all money spent by all participants in travel, accommodations, and other expenses is at least $1,000,000!) For many years the aim was to make each meeting self-liquidating, the Society to pick up the tabs for the occasional shortfalls. Since the middle 1970s the Council has ruled that annual meetings are to be operated so as to produce modest surpluses for addition to the Society treasury.

Until 1925 each Council chose its own meeting place. This was because even then it was held that no Council could take any action to commit any future Council. The 1925 Council chose to break with tradition and authorized an annual meeting in New York in 1928, three years in advance. This action was needed because the 1928 meeting was to be a special one in conjunction with the AAAS. With the ice broken, times and places of virtually all later meetings were planned and authorized a year or more ahead. In recent years invitations from potential host groups are accepted and plans organized from five to eight years ahead. Meeting places that can adequately accommodate conventions of from 3,000 to 5,000 attendees are scarce, and facilities are booked so far ahead that long forward planning is essential.

In the early years the local host organizations carried the costs of annual meetings themselves, except for travel and accommodations which were of course paid by participants or their employers. Not until 1932 were travel fares allowed to officers and councilors, and even then they paid part or all of their own other expenses.

In 1930 the Council first authorized collection of $2 registration fees from all attendees in order to lessen the burden on the hosts. Such fees have been prescribed for nearly every meeting since, and they have risen markedly with the times. Though registration fees have been high enough in recent years to generate periodic complaints, they represent such a small fraction of the total expenses of each individual participant that they seem immaterial.

The Council retains responsibility for accepting invitations from local groups who wish to host meetings; it also ratifies all plans and budgets prepared by the local committees. Other than this mild control, production of an annual meeting is a joint effort by the Local Committee, the headquarters staff, and the Joint Technical Program Committee. This last is a large group composed of representatives from GSA itself, the Local Committee, all the associated societies, and each of the GSA divisions. Its sole function is to select abstracts for oral or poster presentation, to arrange the scientific program, and to bear the brunt of adverse criticism about the multiplicity of papers and sessions that finally appear on the program.

The Local Committee and the general chairman for each annual meeting are drawn from among the host organizations. The committee is activated as soon as a site for the meeting is selected by the Council. Hence it serves for several years, its activities gradually increasing to crescendo proportions during the week-long span of the meeting itself. Service on an annual meeting committee is voluntary and is considered something of an honor. It is deeply appreciated by the associated guest societies, both in writing and by public acknowledgement during the banquet. Service also requires much work over long periods, which perhaps explains why the same faces seldom appear on a local committee more than once. Local committees are divided into a number of subcommittees, each responsible for a single facet of the complex overall meeting—housing, social events, technical meetings, field trips, and the like.

For many years, and except for a couple of New York meetings when GSA headquarters acted as one of the hosts, the organization, management, and financing of each annual meeting were left entirely to the Local Committee and the host institutions. Headquarters and the Council offered helping hands and advice when needed, and it commonly took care of printing and distribution of programs and menus, but in general it kept hands off.

As time went on, GSA headquarters took larger and larger responsibilities in planning and conducting annual meetings. The change was gradual over many years and almost without fanfare or evidence of a conscious decision to assume greater responsibilities. By the mid-1960s, however, the process was well advanced, and since 1967, after the Society's move to Boulder, an annual meetings manager has been a permanent employee. The work of his (or her) small staff is augmented by a large proportion of the total headquarters staff and by many temporary employees just before and during each meeting.

Part of the reason for headquarters management is financial. Handling, safekeeping, and accounting for all the income and outgo of money of an annual meeting is a major responsibility; establishment of separate bank accounts by local committees in itself raised many questions from the Society's auditors and accountants, and answers to those questions always cost additional fees. Adoption of the concept of preregistration for meetings, which is to everyone's advantage, made money-handling by headquarters virtually mandatory.

Aside from money-management problems, use of experienced professional meetings managers tends toward greater efficiency in all the myriad and often overlapping and conflicting operations involved in producing an annual meeting. Contracts and negotiations with hotels and convention centers, collection of registration fees, sale of exhibit space, assignments of space for technical and social gatherings, preparation and stuffing of registration packets, and printing of tickets for dozens of functions are but samples of the manifold behind-the-scenes activities that go into the making of a successful meeting.

The gradual takeover of meeting responsibilities by headquarters has not always been smooth by any means. No easy solutions are evident but more time and more experience will inevitably lead to better mutual understanding of which parts of the complex machine can be better operated by local groups and which by headquarters. Meanwhile, much of the smooth running of annual meetings with minimal irritations and mixups for the attendees, can be attributed to the year-round meticulous devotion to duty and applications of special skills by the annual meetings group at Society headquarters.

Anniversary Meetings

The 25th, 50th, and 75th anniversaries of the Society were special occasions. Planning for an elaborate celebration to mark the 100th anniversary in 1988 was well under way when this history went to press.

The 25th anniversary meeting was held at Yale University, New Haven, in 1912. More than the usual number of scientific papers were given, including 11 broad ones in a general session, 66 more specialized papers grouped in multiple sessions, and 17 papers on vertebrate and invertebrate paleontology. In addition, an evening meeting commemorating the centennial of James Dwight Dana's birth highlighted the conclave. Addresses about Dana's life and works were supplemented by displays of his memorabilia (Council Minutes for 1912, p. 288).

The semi-centennial meeting was in New York in 1938, and it and the preceding meeting in Washington, which also took notice of the approaching birthday celebration, each attracted more than 1,000 people, almost double the usual number of attendees; both were three-day affairs, a day longer than usual. Special efforts to bring geology to the attention of the general public were made through radio broadcasts by several leaders in the geologic profession and by rearranging the usual evening functions to provide time for a well-attended public meeting with popular subject matter. The eight still-living founders and the Honorary Fellows were invited to attend as guests of the Society, but most of them were by then too old or infirm to attend. Publication of a special volume to commemorate the anniversary was authorized. This was intended to be the biography of R.A.F. Penrose, Jr. (Fairbanks and Berkey, 1952), but it could not possibly have been readied in time. Instead, Secretary Berkey managed to assemble a group of 21 papers each reviewing the state of the art in various facets of geology over the period 1888–1938 (Berkey, edi-

tor, 1941). In addition, the scientific papers presented at the meeting were published as a group in the March 1939 issue of the *Bulletin*. The Society paid for all meeting costs of the 50th anniversary, a total of $2,000. No registration fees were charged.

The 75th anniversary was again celebrated in New York in 1963, under the leadership of President Harry H. Hess and General Chairman Robert E. King. Three symposia were featured, and each attracted large audiences from the much larger than usual group of registrants. Subjects were (1) the relation of geology and trace elements to nutrition, (2) ocean basins and continental drift, and (3) uniformitarian concepts in the framework of geological science. Organizers of three symposia were Helen L. Cannon, Harry H. Hess, and Claude C. Albritton, Jr., respectively.

Material presented on geology and nutrition was later published as GSA Special Paper 90 (Cannon and Davidson, 1967, and material on uniformitarian concepts was published as GSA Special Paper 89 (Albritton, 1967).

Albritton was also asked to organize and edit a special anniversary volume on the philosophy of geology. *The Fabric of Geology* (Albritton, 1963) was the result. It was not published by the Society but by a commercial publisher through arrangements with W. H. Freeman, a devoted Society Fellow. Royalties from sales were paid to GSA for addition to its endowment funds. A Spanish translation of *Fabric of Geology* was published in 1970 under the title *Filosofia de la Geologia* by Campania Editorial Continental.

The Geological Society of America will become 100 years old in December 1988. The birthday celebration itself will center around a very special annual meeting, planning for which was begun on a relatively small scale in 1979 and 1980. Far more ambitious and significant than the ceremonial birthday party, however, were plans for a Decade of North American Geology (D-NAG). The decade, which will actually last for something like eight to ten years (closing date is unpredictable now) took firm form in mid-1980 with employment of Allison R. Palmer as Centennial Science Program Coordinator.

The main thrust of D-NAG is described as follows:

The geological community of the continent is being invited to participate in a grand synthesis of the regional geology and geophysics of North America and adjacent oceanic regions. As a stimulus for future research, this synthesis will bring into focus by means of maps, charts, transects, and regional memoirs—including summary volumes on the geology of North America—the wealth of new information about the North American Plate which has been acquired during the plate-tectonics revolution. [*GSA News & Information,* v. 2, no. 9, p. 129. More detailed descriptions of the scheme and its parts appear in *GSA News & Information,* v. 2, no. 11, p. 165, and in later issues of the newsletter.]

The large number of synthesis volumes on all parts of the North American continent will be supplemented by other volumes, by maps and charts on specialties within geology, by papers or symposia at annual, sectional, and divisional meetings, and by many other means.

Initially, D-NAG was to be financed by advances from the Society's Reserve Fund, but success or failure of the entire program would appear to depend heavily on successful fund-raising efforts of the GSA Foundation, established in 1980. Estimated total needs for D-NAG over the 1980–1990 period are about $4 million. The project is sponsored by GSA, but it is to involve the work of several hundred individual authors and numerous institutions. Some of the books will be published by the Society, others by the Geological Survey of Canada; other products will be published by national geological surveys or by other societies.

Nontechnical and Social Functions

Many activities other than presentation of technical papers characterize each annual meeting. GSA and most of the associated societies take advantage of the gathering to hold their council or other business meetings, their awards ceremonies, and their luncheons and banquets. Many geology departments hold receptions for alumni and friends. Scores of manufacturers, publishers, and institutions reach potential customers and friends with outstanding (and costly) exhibits. Job-seekers meet employee-seekers. Short courses, geologic field trips, and excursions for nongeologic attendees are regularly on the program, either during the regular meeting or before or after it. Not since 1960 has the parent Society taken responsibility for organizing field trips or the guidebooks for them, but no local committees can ever resist the opportunity to show off their own geologic environments, hence they conduct numerous trips before, after, and even during each meeting. Some of the major, well-established events are summarized here.

Corporate Meeting. Like any other corporation, The Geological Society of America, Inc., must have an Annual Corporate Meeting. Except for the war years of 1942–44, when annual scientific meetings were not held and the required business meetings took place at New York headquarters, the corporate meeting is held in conjunction with the annual meeting. The business conducted at the corporate meeting is again like that of other corporations— election of officers, appointment of auditors, and ratification by the membership (stockholders) of all actions of the Council (board of directors) during the past year. Unlike most corporations, however, there are no oral reports from officers as to the state of the corporation and seldom any discussion or questions from the membership. This is because virtually all voting is done by written ballots based on summary reports distributed well in advance of the meeting. The use of proxies for voting at the annual meetings was early recognized as essential because of the continent-wide distribution of the membership. It was first authorized in the first revision of the Constitution, adopted at the second annual meeting, December 28, 1889 (Fairchild, 1932, p. 94), though mail ballots were recognized even earlier. Provision is made, of course, for members to vote in person, to withdraw their proxies, or to raise new items of business during the corporate meeting, but these privileges are seldom used.

All participants in corporate meetings follow a rigid script prescribed by Society lawyers. Meetings are held early, before the technical sessions for the chosen day. Aside from a handful of curious members, attendees are commonly limited to those who have been assigned speaking parts in the script, with necessary backups such as attorney, notary public, and two or three staff members to record and guide the solemn proceedings. The entire meeting seldom if ever requires more than a half an hour. Despite its routine nature, the annual corporate meeting is a necessary one, required both by the Society Constitution and by corporate law. The massive disinterest in the business meeting and the management of the Society exhibited by the rank and file seems unfortunate, but it is unlikely to change.

Presidential Addresses. Addresses by the outgoing presidents have been regular features of the annual meetings since the beginning. Decisions as to subject matter, time, place, and duration are made by the President himself. Most addresses have dealt with a geologic subject of special interest to the speaker, whether a summary of his life work or of some recent special research; a few have emphasized the health of the Society or geology in general. One presidential address (Leopold, 1973) was presented partly in song with guitar accompaniment. Preparation of the original music for reproduction in the *Bulletin* constituted a special but not insurmountable problem for the publications manager.

Presidential addresses have been given before or after the annual dinner, or at other times during the technical meetings, commonly without competition from concurrent sessions.

Nearly all addresses have been published in the *Bulletin.* Addresses are commonly considered to be privileged communications, with no discussion at time of delivery and no peer review of presidential manuscripts. Very rarely, a President may permit publication of a discussion of his address, but the decision is his to make.

The Annual Dinner. The custom of holding an annual dinner was established at the Columbus, Ohio, meeting in 1891, when 23 Fellows and three guests shared in a repast that has since become the outstanding social event of each annual meeting. As described rather fully by Fairchild (1932, p. 205), the early dinners were mainly sentimental affairs, full of self-congratulatory reminiscence about the trials, tribulations, and visions of the Society's founders. Beginning about 1920, when attendance was greatly enlarged by GSA's growth and by members of the associated societies, the dinners became much more formal and rather rigidly scheduled.

Attendance at annual dinners ranges widely, seemingly more directly dependent on price and on other attractions of the host city than on factors such as overall meeting registration. In recent years attendance has ranged from 600 to 1,600, always larger than any social or technical function except the welcoming party.

For many years, presentations of the Penrose and Day medals, plus occasional other awards, have highlighted the after-dinner program. The outgoing President, who is the chief guest of honor other than the medalists, commonly gives an account of his stewardship, introduces his successor, and may even give his presidential address unless he has chosen to give it elsewhere during the annual meeting.

Seating arrangements at the head table and for a few of the tables just below it follow strict rules of protocol, planned by the staff in accordance with the President's wishes. The President (or his wife) stipulates the dress code for the head table. The entire affair is designed to allow the Society to do honor to the President and to all others who have been responsible for the Society's welfare during the past year.

The dinner is commonly preceded by two receptions. One is a relatively small invitational affair for medalists and other honored guests and for special friends of the outgoing President and his wife. Personal invitations come from the presidential couple, who also take leading parts in arranging the decor and refreshments offered. The other much larger reception is open to "all others" who are attending the dinner. It commonly features cash bars and snacks; background music or other entertainment may be provided. Mild complaints have risen from time to time about the two-class system of entertainment, but no change seems imminent.

Smoker—Welcoming Party. "In earlier years this (the smoker) was a casual and unpremeditated event. . . .The first mention appears to be in Volume 14, page 534 (of the *Bulletin*), following the presidential address of N. H. Winchell at the Washington meeting in 1902. Since 1918 the smoker has received bold notice in the accounts of the meetings. . . .as occurring after the presidential address in an evening session" (Fairchild, 1932, p. 208).

The general character of the smoker remained almost constant for many years. Usually held after the presidential address (until this event became part of the annual banquet), the smoker featured free cigars (later cigarettes), pretzels, and draft beer, innumerable concurrent conversations, renewals of old friendships, and making of new ones.

In more recent years, the term "smoker" has been dropped and the event changed to a welcoming party, held the night before commencement of the Society's technical meetings. These parties, known at times as "attitude adjustment"gatherings, are commonly characterized by cash bars, appetizers of various kinds and quantities, and musical or other entertainment. As one of the outstanding social events of each annual meeting, they now attract several thousand people, but the chief objective—that of offering a chance to see and be seen—is unchanged. Welcoming parties have been held in headquarters hotels, museums, river steamers, and other locales. They are invariably pronounced as successful, particularly by those with well-developed herd instincts.

During at least one period in Society history, spontaneous group-singing characterized the latter parts of the smoker evenings. On objections from some of the more conservative members, the custom was stopped by Council action in 1952, but a separate beer-room was provided for those moved to song.

Section Meetings

Annual meetings of the six sections and the problems of producing them are smaller replicas of the Society's annual meetings. They are smaller in every way, though several of them are now far larger in attendance than the main Society meetings were only a few years ago. The Council generally sends the President or his delegate to each sectional meeting as evidence of the parent's interest in its offspring, and the Council approves the general plans for each meeting. The Society also publishes the abstracts and programs for each section. Otherwise the production of a section meeting is done almost entirely by its local committee. The meetings are all intended to be self-liquidating; money is handled by the section treasurers. During the 1970s, problems arose as to financial accounting and possible tax consequences of local banking of what were legally GSA funds. These led to decisions by the Council that financial management should be handled by headquarters. This in turn led to unhappiness on the parts of several sections. Finally, in 1978 the Council changed the Bylaws to permit the sections to incorporate separately if they so wished, with freedom to manage all of their own affairs. When this book went to press, none of the sections had chosen to incorporate.

12

Honors and Awards

Introduction

Like most societies, learned or otherwise, the Geological Society of America presents various honors and awards to individuals for outstanding accomplishments in the earth sciences. By invitation, the Society also recommends one or more earth scientists for consideration for the National Medal of Science, awarded annually by the President of the United States.

The Society's most prestigious awards are the Penrose Medal and the Arthur L. Day Medal. Certain internationally outstanding geologists are also elected as Honorary Fellows. Elections as officers or councilors, or appointment to committees, all are also considered honors by most recipients. In addition to these Society-wide awards, several of the divisions give awards to individuals within their specialties. Election or appointment to Society service excepted, none of the awards are restricted as to residence or Soceity affiliations of the recipients. They are thus international honors.

Correspondents (Honorary Fellows)

The original Constitution provided for election of foreign Correspondents; however, no Correspondents were actually elected until 1909. The Council Minutes for October 14, 1909 (as quoted by Fairchild, 1932, p. 107), explain the long delay thus: "Appreciating the fact that the Society is now twenty-one years old, and feeling that the organization has attained sufficient standing to render election to corresponding membership in it an honor to be prized, the Council recommends to the Fellowship the following seven men for election to Corresponding Members. . . ."

Since 1909 and except for a few war years, one or more Correspondents (now called Honorary Fellows) have been elected annually. The numbers in any one year have ranged as high as seven, but more commonly only one or two are chosen. The total number of Correspondents—Honorary Fellows living at any one time has seldom if ever exceeded 60. The list is too long to give here, but up-to-date lists of living Honorary Fellows appear regularly in the *Membership Directory* and names of all former honorees can be found in the annual reports and other sources.

Geologists who have distinguished themselves in geological investigations or in signal service to the Society may be elected as Honorary Fellows. In practice, nearly all candidates have lived and worked elsewhere than North America. The most noteworthy exceptions were astronauts. Caught up in the fervor of the Moon landings in 1969, the Council quickly amended the Bylaws to permit North Americans to become Honorary Fellows under exceptional circumstances. The Apollo II astronauts who had first walked on the Moon—Neil A. Armstrong, Edwin A. Aldrin, Jr., and Michael Collins, were elected; they were feted at the annual meeting in Atlantic City. In 1973 another visitor to the Moon and the only professional geologist to have done so, Harrison H. Schmitt, was elected to Honorary Fellowship.

Most Honorary Fellows have been elected toward the evenings of their careers, after lifetimes of outstanding and internationally recognized contributions to the science. Aside from the honor itself, perquisites are few—a letter from the President, lifetime remission of dues with publications and other privileges as available to the membership, and introduction at the annual banquet or other function. the Society may, but seldom does, pay travel expenses for first-time attendance of a newly elected Honorary Fellow.

Penrose Medal

The Penrose Medal, generally considered to be the Society's highest honor for achievement in the geologic sciences, was established at the end of 1925, when the Council accepted an offer from R.A.F. Penrose, Jr., to endow such a medal. A few years earlier Penrose had presented a similar medal to the Society of Economic Geologists, which he had helped found and had served as its first president. Penrose called his gift simply the Geological Society of America Medal, but not surprisingly it soon became known as the Penrose Medal, its official name ever since.

By specification of the donor, the Penrose Medal is awarded in recognition of eminent research in pure geology—outstanding contributions or achievements that mark a decided advance in the science of geology. In consequence, the list of Penrose medalists, given below, contains names of the fathers of many of the concepts on which modern geologic knowledge is based.

The Penrose Medal was designed by Penrose himself, though he worked with an artist to transform his conceptual sketches into final form. His original sketches remain on file in Boulder. With remarkable prescience, Penrose foresaw the day when geology would reach far beyond the planet Earth. The obverse of his medal, showing the earth and moon as one would view them from outer space, was intended by the donor to be astronomically and topographically correct. With his use of stylized fossils, of a multi-vent volcano, and other features—and by his use of gold as a medium—he condensed most of the grand sweep of earth sciences on one small disk.

The medal is of 14-karat gold, weighs 3 ounces, and is 2¼ inches in diameter; a bronze copy is also given to each recipient. Another bronze copy is on display in the conference room at GSA headquarters, on a plaque bearing the engraved names of all recipients. A companion plaque shows the Day Medal. Election to lifetime fellowship accompanies the awards of Penrose and Day medals. Monetary awards are never given: this was a firm stipulation by the donor of the Day Medal and is a rigid custom for the Penrose Medal also.

Awards of the two gold medals are traditional highlights of the Society banquets at annual meetings. Remarks made during presentation and acceptance of the medals are regularly published in the *Bulletin* or in other Society publications.

To provide funds for his medal, Penrose gave $5,000 in City of Philadelphia bonds, which yielded, at 4%, $200 per year. With gold at $20.67 per ounce, this yield was more than adequate to purchase medals with three ounces of 14K gold and to pay travel expenses of the recipients. Astute as he was, Penrose could not have foreseen that gold would rise to $35 per ounce within a short time after his death, much less that it would far exceed $700 per ounce during the first fortnight of 1980—and would remain at or above $500 thereafter.

The original gift, with its constant income of $200, remained intact until 1947. Because the issue was soon due for redemption, the Philadelphia bonds were then replaced by Northern Pacific Railroad 4½% bonds. Since that time, the Penrose Medal Fund has remained segregated from other accounts and has been invested in various media from time to time. In 1978 the principal stood at only $3,642 with an income that was clearly insufficient to cover the costs of striking and presenting the medal.

In 1979 the Council was still wrestling with the problem of shortfall of income for the Penrose Medal. The interim solution was to reduce transportation allowances for the medalists and to lay in a supply of several medals so as to postpone the effects of further inflation in gold prices for a few years. The Council also transferred additional money from the Current Fund to the Penrose Medal Fund, much as it did for the Henry R. Aldrich Fund when it was ailing (Chapter 16).

Other possible avenues for relief seem available and will no doubt be considered by future Councils. Discontinuance of the custom of paying travel and living expenses for the medalists would save several hundred dollars per year; it would, moreover, fit with Penrose's own wishes. In 1930 and while he was Society President, he told the Council that he did not think it advisable to pay travel expenses of medalists from the fund he had established. A cut in the size of the medal or reduction of the fineness of the gold used in it would be helpful but would doubtless be considered contrary to the wishes of the donor. (In his letter of endowment, however, Penrose emphasized that he had no objections to changes in the conditions of the award or in design of the medal, as future Councils might see fit.)

One extreme solution has already been considered and rejected. The medalist for 1945, F. A. Vening-Meinesz, asked that he be given a bronze medal and that the money saved go to some needy professor in war-devastated Europe! On advice from its legal counsel that a change from gold to bronze would be a breach of faith with the donor, the Council refused the offer with thanks.

Arthur L. Day Medal

The Day Medal was endowed in 1948 by Dr. Arthur L. Day, long-time leader of the Geophysical Laboratory, Carnegie Institute, in Washington. As had Penrose, Dr. Day provided the die for the medal plus $7,000 to fund the endowment. As with the Penrose award, the medal was

Penrose Medal

 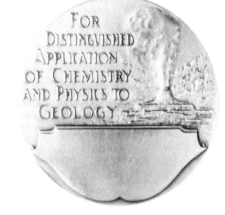

Day Medal

specified to contain 3 ounces of gold, not less than 14 karats fine. Unlike the Penrose Medal, which was designed to recognize a near-lifetime of eminent research in pure geology, the Day Medal was "to provide a token which shall recognize outstanding achievement and to inspire further effort, rather than to reward a distinguished career." Moreover, the medal is intended only for those who have applied physics and chemistry to the solution of geologic problems. In consonance with the donor's wishes, the Day Medal goes annually to a scientist who is younger than most Penrose medalists, but who has already had outstanding accomplishments in the fields of geochemistry or geophysics.

The medal was designed mainly in collaboration between Dr. Day and Secretary Henry Aldrich.

The fund established by Dr. Day has fared much better through the years than the Penrose Medal Fund. It was larger in the first place—$7,000 versus $5,000 for the Penrose Medal—and it has been differently invested. In consequence, the Day Medal Fund stood at $50,567 in 1978, and provided income of $598 (Treasurer's Report, 1978). Use of the excess money in the Day Fund has probably caused more anguish on the part of Councils and headquarters staffs than has the shortage in the Penrose Medal Fund. Dr. Day originally wished any excess income to be used to buy life memberships for the recipients. Since all recipients of the Penrose and Day medals now automatically become life members with remission of dues, this provision is no longer applicable. Nowadays, excesses of income from the fund are either allowed to accumulate or in tight-budget years are added to other moneys available for research grants.

Penrose and Day Medals Contrasted

The evolution of the Penrose and Day medals is a study in contrasts. Penrose designated his gift as the Geological Society of America Medal; the name Penrose Medal was not adopted until after his death. He designed the medal himself and had it drawn in final form by an artist. He had the dies cut and the first copy of the medal struck by the United States Mint in Philadelphia. The terms of his endowment for the medal and definition of his wishes were so clearly and concisely drawn that no questions of interpretation have ever plagued successive Councils.

Guidelines for the casting and award of the Day Medal, on the other hand, required a full year to complete, after confusion marked by much talk, innumerable letters, and several trips to Washington by the Society secretary and others to discuss things with Dr. Day in person. The purpose of the medal, its design, whether it was to bear the donor's name, whether it was to be of gold or bronze, the terms of the endowment (particularly as to Dr. Day's wishes should income from the endowment ever result in

surplus funds)—all these and more problems had to be settled before the proffered gift became reality. Long after the donor's death, some problems continued to arise. Mrs. Day could answer some of these from her memory of her husband's wishes, but as late as the early 1970s, there were still some questions as to permissible uses of surplus funds.

The donor's lack of clarity in thinking out exactly what he wished to accomplish with his gift must be blamed for most of the problems that arose. Others were caused by honest misunderstandings and by Dr. Day's initial insistence that his plan to endow a medal was to be kept as a private matter between him and Secretary Aldrich, forcing Aldrich to forego open guidance and advice from others. Enough of the confusion was finally resolved, however, to permit striking of the first Day Medal just in time for its presentation to George W. Morey at the 1948 annual meeting. This was fortunate, for Morey had already been named the recipient in anticipation of the gift, but the medal he was to receive was not available until the last possible moment.

Division Awards

Kirk Bryan Award

The Kirk Bryan Award, administered by the Quarternary Geology and Geomorphology Division, was established in 1951, but was not actually presented until 1958. Generally given annually, the Bryan Award goes to the author or authors of a published paper that marks distinct advance in the science of geomorphology or a related field. The award itself, commonly presented at a division luncheon during the Society's annual meeting, consists of a certificate and a cash stipend. It is financed by income from the Kirk Bryan Memorial Fund which was, in turn, financed by contributions from many friends of Kirk Bryan.

The certificate for the Kirk Bryan Award itself records an interesting bit of history (interesting, at least, to those who knew R. C. Moore). The proposers had much trouble and delay in designing a certificate that was acceptable to both the division's officers and to the GSA Council. In hopes of breaking the deadlock, then GSA President Raymond C. Moore worked up a substitute design himself. He asked the Society's engraving company to set his design for printing and to produce proofs that he could show to the Council. By mistake (?) the printer went ahead to print an entire edition! Not surprisingly, the Council and the Division approved and adopted the Moore design (Council Minutes for 1958, p. 56).

Since 1974 the division has also sponsored a research grants program in memory of J. Hoover Mackin. Consisting of a cash grant in support of research, it is presented annually as part of the Society's research grants program.

Beginning in 1982, the Division on Quaternary Geol-

PENROSE MEDALISTS

1927	Thomas Chrowder Chamberlin	1955	Maurice Gignoux
1928	Jakob Johannes Sederholm	1956	Arthur Holmes
1929	*No award given*	1957	Bruno Sander
1930	François Alfred Antoine Lacroix	1958	James Gilluly
1931	William Morris Davis	1959	Adolph Knopf
1932	Edward Oscar Ulrich	1960	Walter Herman Bucher
1933	Waldemar Lindgren	1961	Philip Henry Kuenen
1934	Charles Schuchert	1962	Alfred Sherwood Romer
1935	Reginald Aldworth Daly	1963	William Walden Rubey
1936	Arthur Philemon Coleman	1964	Donnel Foster Hewett
1937	*No award given*	1965	Philip Burke King
1938	Andrew Cowper Lawson	1966	Harry H. Hess
1939	William Berryman Scott	1967	Herbert Harold Read
1940	Nelson Horatio Darton	1968	J. Tuzo Wilson
1941	Norman Levi Bowen	1969	Francis Birch
1942	Charles Kenneth Leith	1970	Ralph Alger Bagnold
1943	*No award given*	1971	Marshall Kay
1944	Bailey Willis	1972	Wilmot H. Bradley
1945	Felix Andries Vening-Meinesz	1973	M. King Hubbert
1946	T. Wayland Vaughan	1974	William Maurice Ewing
1947	Arthur Louis Day	1975	Francis J. Pettijohn
1948	Hans Cloos	1976	Preston Cloud
1949	Wendell P. Woodring	1977	Robert P. Sharp
1950	Morley Evans Wilson	1978	Robert M. Garrels
1951	Pentti Eskola	1979	J Harlen·Bretz
1952	George Gaylord Simpson	1980	Hollis D. Hedberg
1953	Esper S. Larsen, Jr.		
1954	Arthur Francis Buddington		

ARTHUR L. DAY MEDALISTS

1948	George W. Morey	1965	Walter H. Munk
1949	William Maurice Ewing	1966	Robert M. Garrels
1950	Francis Birch	1967	O. Frank Tuttle
1951	Martin J. Buerger	1968	Frederick J. Vine
1952	Sterling Hendricks	1969	Harold C. Urey
1953	John F. Schairer	1970	Gerald J. Wasserburg
1954	Marion King Hubbert	1971	Hans P. Eugster
1955	Earl Ingerson	1972	Frank Press
1956	Alfred O. C. Nier	1973	David T. Griggs
1957	Hugo Benioff	1974	A. E. Ringwood
1958	John Verhoogen	1975	Allan Cox
1959	Sir Edward C. Bullard	1976	Hans Ramberg
1960	Konrad B. Krauskopf	1977	Akiho Miyashiro
1961	Willard F. Libby	1978	Samuel Epstein
1962	Hatten Schuyler Yoder	1979	Walter M. Elsasser
1963	Keith Edward Bullen	1980	Henry G. Thode
1964	James Burleigh Thompson, Jr.		

RECIPIENTS OF THE KIRK BRYAN AWARD

1958 Luna B. Leopold, Thomas J. Maddock, Jr.
1959 Jack L. Hough
1960 John F. Nye
1961 John T. Hack
1962 Anders Rapp
1963 Arthur H. Lachenbruch
1964 Robert P. Sharp
1965 Gerald M. Richmond
1966 Charles S. Denny
1967 Clyde A. Wahrhaftig
1968 David M. Hopkins
1969 Ronald L. Shreve
1970 Harold E. Malde

1971 A. Lincoln Washburn
1972 Dwight R. Crandell
1973 John T. Andrews
1974 Robert V. Ruhe
1975 James B. Benedict
1976 Geoffrey S. Boulton
1977 Michael Church
1978 Richard L. Hay
1979 Stanley A. Schumm
1980 James A. Clark, William E. Farrell
and W. R. Peltier

RECIPIENTS OF THE E. B. BURWELL, JR., AWARD

1969 Lloyd B. Underwood
1970 Glenn R. Scott, David J. Varnes
1971 Edwin B. Eckel
1972 R. J. Proctor
1973 J. E. Hackett, Murray R. McComas
1974 Robert F. Legget
1975 Erhard M. Winkler

1976 David J. Varnes
1977 Richard E. Goodman
1978 Nicholas R. Barton
1979 John W. Bray, Evert Hoek
1980 Kerry E. Sieh

RECIPIENTS OF THE GILBERT H. CADY AWARD
(The award generally is made biennially.)

1973 James M. Schopf
1974 *No award given*
1975 Jack A. Simon
1976 *No award given*
1977 William Spackman, Jr.

1978 *No award given*
1979 Peter A. Hacquebard
1980 *No award given*

RECIPIENTS OF THE O. E. MEINZER AWARD

1965 Jozsef Toth
1966 Charles L. McGuinnes
1967 Robert W. Stallman
1968 Mahdi S. Hantush
1969 Hilton H. Cooper, Jr.
1970 Victor T. Stringfield
1971 George B. Maxey
1972 Joseph F. Poland, George H. Davis
1973 William Back, Bruce B. Hanshaw

1974 R. Allan Freeze
1975 John D. Bredehoeft, George F. Pinder
1976 Shlomo P. Neuman, Paul A. Witherspoon
1977 Jacob Rubin, Ronald V. James
1978 R. William Nelson
1979 Patrick A. Domenico, John M. Sharp, Jr.
1980 Richard L. Cooley

ogy and Geomorphology will be expected to advise the Committee on Research Grants on yet another research grant—the newly established Gladys W. Cole Memorial Research Award. This award, established in 1980 by W. Storrs Cole in memory of his wife, has several other restrictions, but it is intended primarily to apply the fund's interest for research on geomorphology of the arid lands in Mexico or the United States (*GSA News & Information*, v. 2, no. 12, p. 183).

E. B. Burwell, Jr., Award

The Burwell Award was established in 1968 by the Engineering Geology Division to honor the memory of Edward B. Burwell, Jr., one of the founders of the division and long the Chief Geologist, U.S. Army Corps of Engineers. The award is given annually to the author or authors of a published paper of distinction in the field of engineering geology, or of related fields of applied soil or rock mechanics. It consists of a certificate and a modest cash award and is presented at the division's annual business luncheon. Costs are provided from annual dues collected from division affiliates.

O. E. Meinzer Award

The Meinzer Award was established by the Hydrogeology Division in 1963 in memory of Oscar E. Meinzer, long known as the father of ground-water geology in North America. The award is presented annually to the author or authors of an outstanding published paper advancing the science of hydrogeology or some related field. The award consists only of a certificate and temporary retention of the Birdsall bowl, described separately below. It is commonly presented at the annual luncheon meeting of the division.

Birdsall Bowl and Bequest. Not long after the O. E. Meinzer Award was established in 1963, John M. Birdsall, banker, businessman, and long-time ground-water geologist with the U.S. Geological Survey, presented to the Hydrogeology Division a large and impressive sterling silver Revere bowl. The Birdsall bowl has remained as the property of the Division and since 1967 has been presented annually to the winner of the O. E. Meinzer Award. Awardees' names are engraved on the bowl, and each is

allowed to retain it for one year if he so wishes, when it is returned to headquarters for engraving of an additional name and presentation to the next awardee. In 1977 the division voted to broaden the award and to add immeasurably to its sentimental impact by presenting a smaller replica of the Birdsall bowl to each recipient for his permanent retention. At the 1978 annual meeting the new scheme was instituted, with presentation of replicas to all former living awardees and to the widows of three deceased winners.

Recurrent expenses connected with the Birdsall bowl awards—engraving, refurbishing, transportation, and replicas—are paid from income from the Birdsall Bequest. When Birdsall died September 11, 1975, he generously bequeathed $10,000 in cash to the Hydrogeology Division. Though the Birdsall bequest fund is managed like all other Society investments, its use is determined by the division (decisions are ratified by the GSA Council).

The Birdsall bequest yields more income than is needed for care of the bowl and provision of replicas. The remainder of the income is used for Birdsall Distinguished Lecturer tours. Established by the division in November 1976, these consist of a series of lectures at one or more universities by an outstanding hydrogeologist. The first series was given in 1977 by Jacob Bear at the University of Wisconsin. The second, in 1978, was by William Back and involved several lectures on geochemistry of ground water at a number of widely scattered universities. The lecture series has been unqualifiedly successful and has already proved to be a wise use of Birdsall's gift. His memorial appears in La Moreaux and Barksdale (1977).

Gilbert H. Cady Award

The Cady Award was established by the Coal Geology Division in 1971 and was named in honor of Gilbert H. Cady, one of America's foremost authorities on the geology of coal. It is financed by income from the Gilbert H. Cady Memorial Fund, to which division members and others of Cady's friends contributed. The award is made for outstanding contributions (not necessarily in the form of published papers) in the field of North American coal geology. Given biennially, it consists of a certificate and a handsome sterling silver tray.

13

Penrose Conferences

The Penrose Conferences, established in 1969, have already come to be among the most fruitful methods of communication between geologists that have been conceived during the past half century or more. In the decade 1969–1979, more than 60 conferences were held, all but a handful fully meeting the objectives of their conveners and of the Society. With average attendance of about 60 people and even with some duplication, this means that nearly 4,000 scientists have taken part directly. Each of these has shared his new knowledge with colleagues, thus helping spread awareness of facts or theories both in North America and abroad to an incalculable degree.

The Penrose Conferences, so named as one more tribute to GSA's prime benefactor, were modeled on the Gordon Conferences, well known in the chemical profession for many years. Extension of the Gordon Conference concept to geology and sponsorship by the Society evolved like almost every new departure in GSA history. So far as can be seen in the Council Minutes, the seed was planted independently but concurrently by Howard J. Pincus and Robert F. Legget in the fall of 1966 in letters and reports to the Council. The 1967 Council approved establishment of the conference series, and for several years Legget took the lead in organizing it.

The first conference, convened by Brian J. Skinner, was held in January 1969 in Tucson, Arizona. The subject was "Depositional Environment of the Porphyry Coppers." This conference was in fact originally organized and scheduled as an activity of the Research Committee of the Society of Economic Geologists, but it coincided with the ripening of GSA's plans to initiate the Penrose series. The convener and the SEG conference management generously

agreed to transfer the auspices to GSA. Since then, from two to as many as ten conferences have been held each year. Attendance has generally ranged between 50 and 70, which is considered the maximum allowable for optimum communication. Most conferences have been in the United States, but some have been in Canada and in Europe. There are no firm restrictions as to place, but the Society tends to frown on non–North American locales because of cost, logistics, and other difficulties.

The purpose of Penrose Conferences is well stated in a recent version of the guidelines (*GSA News & Information*, May 1979, v. 1, no. 5, p. 73):

The Penrose Conferences were established by the Geological Society of America in 1969 as an important effort in its promotion of the Earth sciences. The conferences provide the opportunity for exchange of current information and exciting ideas pertaining to the science of geology and related fields. They are intended to stimulate and enhance individual and collaborative research and to accelerate the advance of the science by the interactions and development of new ideas. The conferences consist of a critical mass of active scientists from the Society, the national and international science communities, and students, sequestered in an attractive meeting place for several days of focused discussion. The participants do not seek simply to resolve technical controversies; their objectives are to provide stimulus and excitement for their field, to air new ideas and develop new associations, and to provoke new research on important questions.

Ideal subjects for conferences are those Earth science topics for which recent work suggests a potential for further significant advances in the near future. Each conference subject should be under current investigation and active discussion by a number of able researchers in the field and/or in the laboratory. Topics should be broad enough so that a range of specialists can discuss them from several points of view, but not so broad that a lack of communication can develop.

Conferences provide an opportunity for a group of like-minded individuals to live, work, and relax together in pleasant (sometimes plush) surroundings, away from the usual distractions of the workaday world; families or casual visitors are not permitted. Most conferences last five days, and many feature field trips to good examples of the subject matter under discussion. Conferences are only lightly structured, as rigid structuring tends to lead to a procession of formal papers, with little discussion or introduction of new ideas. In other words, a fruitful Penrose Conference tends to be like a well-run "brainstorming session."

A set of rules or guidelines, engendered by the Penrose Conferences Committee and the Council, governs the convening and holding of conferences. These have been changed in detail from time to time as experience has accumulated, but most of the basic premises are unchanged since the program's inception. Any individual or group may volunteer to convene a conference so long as at least one convener is a GSA member. Attendees are selected by the convener from those who have expressed interest in participating and from a group of highly recommended students, who are partly subsidized by other participants. Conveners commonly also invite a few key speakers who are known leaders in the conference subject matter. To provide for free discussion of controversial or confidential matter, no scientific reports on conferences are to be published, though summary news reports are required from the conveners, and participants are encouraged to discuss the results with their colleagues. (This restriction has caused complaints in some quarters, mostly from individuals who were not invited to attend; the standard reply is that dissidents are free to propose conferences of their own on the same or different subjects.)

Though the Society carried most of the administrative and some other costs during the formative experimental stages of the program, more recently each conference must be self-supporting. Each participant pays a fee calculated to cover all costs of attendance including overhead. Each also pays his own transportation to and from the meeting. Services needed in planning and operating the conferences, including choice of meeting site and preparation of necessary contracts are provided by a Penrose Conference Coordinator. At first this person was a member of the staff at GSA headquarters, but later the responsibility was turned over to an outside contractor. The selected contractor had resigned from the GSA staff to form her own company for managing technical meetings, so her accumulated knowledge of Penrose Conferences continued to be available. More recent, conveners have been encouraged, but not required, to utilize the services of a professional conference coordinator.

14

Research Grants

Establishment of a modest research grants program was one of the firm recommendations of the Advisory Committee on Policies and Projects, appointed shortly after Penrose's death (Chapter 4). A minor flood of requests for financial aid was received shortly after news of the Penrose bequest reached the membership. Some of these represented genuine pent-up need for assistance in completing research already under way. Some almost certainly reflected personal financial crises during the Depression or the well-known human tendency to reach out eagerly for giveaways.

Oddly, no one on the committee or the Council seems to have been aware that Penrose himself was averse to direct financial support of research as a means of advancing the science; he preferred medals, awards, and publication of research results (Fairbanks and Berkey, 1952, p. 702). The American Philosophical Society, GSA's co-heir, also ignored Penrose's personal preferences, if they were indeed known at the time, and decided to devote much of its Penrose bequest income to support of research. In any event, the Council accepted the committee's recommendation and initiated a grants program late in 1932 that has continued uninterruptedly to the present day, though with minor changes to fit the times.

Grant Recipients

The Advisory Committee and the Council envisioned short-term grants for use in completing or publishing worthwhile research that was already far advanced. Only in exceptional cases were grants to be made to start new pro-jects and then only to experienced workers. These restrictions were then in tune with the times, for the Society consisted only of mature geologists, with no Members or Student Associates; not for many years would grants be opened to those outside the club. This broad policy of aiding only established professionals was followed for many years. In 1933 several ongoing projects of the U.S. Geological Survey were kept alive by means of GSA grants. This was during one of the Survey's darker hours when periods of leave without pay became mandatory for its geologists in order to help the Survey get through the Depression.

Not until 1955 were research grants made available to graduate students, as distinct from practicing professionals, and in support of new projects rather than those that were nearing completion. Even then, only 10% of the grants budget was to be used for Ph.D. candidates. Grants gradually became more numerous but smaller in size, ranging from a few to a couple of thousand dollars, whereas earlier grants had generally ranged upward from $1,000. Between 1955 and 1972 grants were available only to students who were doctoral candidates, and most of them were to support research for dissertations. In 1973 the bars were set even lower and master's degree candidates became eligible. In 1979 the committee seriously studied several requests for support of research by undergraduates; affirmative action was postponed but probably only temporarily.

Funding for the Program

Except for a short period during World War II when non-war-related research was at a standstill and few re-

121

quests for aid were received or funded, the annual totals of research grants have commonly ranged between $20,000 and $125,000. Variations have depended more on the Society's income and economic health than on total need as indicated by number and size of applications for aid. At an estimated average of $50,000 per year, which is probably somewhat low, the Society had devoted more than $2 million in direct support of research.

From time to time, direct appropriations from the treasury have been supplemented by funds from other sources. For example, income from several sizeable gifts from Harold T. Stearns is, by agreement with the donor, regularly used for grants in a particular area of research (see Chapter 12). Similarly, beginning in 1975 a number of gifts, earmarked for addition to the grants program, were received from certain oil companies. These gifts, largely resulting from the personal efforts of a dedicated chairman of the Research Grants Committee, George de Vries Klein, were unrestricted as to use. As a matter of courtesy, however, they were awarded for projects that had some bearing on the donors' interests, and the names of the awardees are reported to the donor organizations.

Periodically, requests have been made to the National Science Foundation for help in funding the research grants program. These have been consistently denied, not for lack of sympathy on NSF's part, but because agency policy prohibits the giving of grants for research projects that it cannot administer and oversee itself. The GSA program, on the other hand, calls for numerous small grants to individual researchers, with virtually no oversight as to use of the money or results of the research.

Until very recently, the Society requested no repayments from grantees after they had completed their educations and entered the work force. In 1975 in line with a general financial tightening, the Council asked former grant recipients to contribute to the research grants fund. Predictably, the response from some was quick and generous, but the total amounts received were not large. Solicitation was dropped in 1977.

Financial and Technical Accounting

Accounting methods applied to grant monies have been greatly changed and simplified through the years. Originally, agreements between the Society and grantees were formal—and formidable—legal contracts that required notarizations and detailed quarterly accounting records and backup vouchers. In some cases the contracts could not be completed or money made available until after the field season for which they were intended. Methods of selection, too, were complicated. Requests for grants were first winnowed by the Committee on Project Grants, then considered and voted on individually by the Council. Progress reports, both financial and technical, were required of

each grantee, and the Society expected scientific papers resulting from the research to be submitted in due course for publication in the *Bulletin.*

In 1947 the old cumbersome system of rigid accounting was greatly simplified. From then on, the researcher was given a lump sum of money; he was honor-bound to use it effectively and economically and to return any unused balance. Only recently have funds been made available promptly and in time for use during summer field seasons. Rather than waiting for Council action on each grant, the Research Grants Committee is authorized to award grants directly up to the limit of the overall amount budgeted for this purpose. Checks are sent to successful applicants immediately after the committee meets in March, and the Council merely ratifies the list of awards at its meeting later in the spring.

Unused balances from research grants have been treated variously. At times they went back to the treasury. At other times they were made available to the committee to finance additional projects, even if this required special meetings of the committee. Nowadays a more streamlined operation permits immediate use of such funds. If a grant is refused or cancelled for any reason, the headquarters staff is authorized to apply the released money to one or more alternate projects that were listed by the committee for this purpose.

Today, brief, informal technical and financial reports are welcomed at the conclusion of research projects, but they are no longer required. Similarly, reports that result from GSA-supported research are not required to be submitted to the Society for possible publication. Naturally, the Society appreciates brief acknowledgement of its support in papers published elsewhere, but even this courtesy is not required.

Changing Policies in the Grants Program

In addition to changes in thought as to whether the Society should aid established professionals or relative beginners in geology, two other questions concerning research grants have arisen time and again. These have to do with large, continuing research projects versus small, short-term grants, and as to whether the Council or individual researchers should identify the kinds of research deserving of support. These were perhaps aired most thoroughly during the period 1936–1940. Initiated mainly by Past President Nevin M. Fenneman, they led to long discussions, bulky reports, and much correspondence by the Council and no less than four committees, active concurrently or at different times. In addition to a Committee of Past Presidents, which was in some respects a forerunner of the Past Presidents, which was in some respects a forerunner of the later Policy and Administration Committee, there were separate research committees having to do with projects, program, and policy. Predictably, as much effort was spent on definitions of committee objectives and duties as on research itself. The story was summarized by Secretary Aldrich in the Council Minutes for 1941 (p. 133–195). Despite the efforts that went into the matter, the results were somewhat inconclusive. The onset of World War II ended the discussion, and the research grants program itself was greatly reduced for the duration. Questions have recurred since the war, but not in as concentrated form as during 1936–1940.

Guidance of Research Directions

From time to time the Committees on Research Grants, backed by the Council, have taken tentative steps to influence the directions of geologic research. For example, not long after the birth of what was to become the famous Paricutin volcano in Mexico in early 1942, the Society set aside $17,500 to support a research program. As it turned out, a major GSA-supported program was not needed, as a much larger one than would have been feasible for the Society developed promptly; it was financed from many sources, both private and governmental. Most of the ensuing studies were arranged through and coordinated by the United States Committee for the Study of Paricutin Volcano, an arm of the National Research Council, and by a parallel Mexican committee.

Though there was never a separate GSA research project on Paricutin, the Society contributed very heavily to the total effort for some years, both during and after the war. Numerous research grants went to individuals, funds were made available for chemical analyses of rocks, and through the American committee, GSA provided for the building of "GSA House" a substantial two-story home and observatory built near the volcano for use by some of the many investigators.

A similar but larger tentative appropriation was made for studies of the effects of the post-war atmospheric nuclear tests on Bikini and other Pacific islands. Again, the project was financed elsewhere and GSA support was not needed. Other suggestions that the Society should take the lead in guiding geologic research in North America have emerged from time to time, and some have been studied at length, but none has ever been followed by positive action.

Long- Versus Short-Term Projects

With a few notable exceptions, the Society has always favored short-term research projects. In most cases grants have been for one year, with extensions or additional funds permitted only on evidence of real progress and promise of early completion. In recent years, any grantee who needs to continue research for more than one year must submit an entirely new application and justify it in competition with all other projects.

Three long-continued research grants, exceptions to the short-term project concept, require mention because of their great contributions to the science; a fourth one is included because it demonstrates a little-known facet of GSA's many-faceted nature.

From 1934 through 1940 the Committee on Projects and the Council regularly granted Robert T. Hill, one of the original Fellows, $1,200 to $1,800 per year to continue his research on the history of early explorations in Texas. The Council Minutes are vague—perhaps intentionally so—as to justification for the project and Hill's progress on it, though some of his findings were published at intervals in the *Dallas News*. Between the lines of the minutes one senses that the continuing grants were in reality a charity to help an aging and distinguished Society member survive intellectually and financially. Such generosity toward an individual, if it was actually so motivated, is rare in GSA history.

One or more members of the 1941 Council must have tired of the small but constant drain on available research funds. The Council voted (Council Minutes for 1941, p. 22) to cut the Hill grant in half and authorized the Secretary to inform Dr. Hill that this would end the support for his project. Hill died shortly after this action, on July 28, 1941, at the age of 83. No product of his research ever emerged. (For his memorial, see Vaughan, 1944.)

One of the three large continuing research projects was that in support of the famous Rock Analysis Laboratory at the University of Minnesota. From 1934 through 1953, the guiding spirit of the laboratory, Professor Frank F. Grout and his successors, received annual grants of $10,000. The total of nearly $200,000 in direct support of the laboratory was small, however, as compared to the indirect support. For many years, and continuing long after the direct grants to Grout ceased, scores of other research grants included items to cover costs of chemical analyses of igneous rocks, the work to be done by Grout's laboratory.

The methods of wet chemical analysis as developed and used at Minnesota have now been largely supplemented or replaced by newer and more sophisticated analytical methods. For many decades, however, Grout's laboratory unquestionably contributed very heavily to geochemistry and the entire science. The Society can well take pride in the very large support that it was able to provide.

A second large research project that GSA supported over a period of years was the revision of the mineralogist's bible—the *System of Mineralogy* by James Dwight Dana (the Society's second President). A complete and painstaking revision and modernization of Dana's system was undertaken in 1936 by W. E. Ford, Charles Palache, and several other Yale and Harvard colleagues. An earlier request for support of a Dana revision had been refused by the 1933 Council. Their efforts resulted in publication of three large volumes (Palache and others, 1944, 1951; Frondel, 1962). Other volumes planned to complete the revision of Dana's *System* have never appeared.

The project was supported financially by grants from Harvard, Yale, the American Philosophical Society, the publisher, and GSA. From 1936 through 1950 the Society appropriated a total of $54,000; of this, $47,350 was actually used by the authors in preparation of volumes I and II of the revision. No GSA support was needed for Volume III and the Society's participation ceased.

Unlike virtually all other research grants, GSA recovered all of the money it had invested in support of the Dana revision project in the form of publisher's royalties. Such repayment was contemplated in the original negotiatins, but no mention of it appeared in the grant documents. Nevertheless, the publisher, John Wiley and Sons, insisted on making good on its informal commitments. The history of the project and its finances, as viewed from the Society's standpoint, is summarized in a long report by Secretary Aldrich to the Council (Council Minutes for 1957, p. 291–301). This research project, then, was one of very few that have been self-liquidating. Even so, the Society can take pride in having made the revision feasible when support was needed most.

Finally, the *Treatise on Invertebrate Paleontology* requires notice here only because it was supported in its early years by repeated research grants, not only to its originator, Dr. Raymond C. Moore, but also to some of the paleontologists who contributed chapters. The story of the *Treatise,* financial and scientific, is a tangled one, and the series is not yet finished. A condensed version of its history appears in Chapter 10 on Publications. Suffice it to say here that from 1948 through 1957, research grants totaling about $40,000 went to Dr. Moore in direct support of the *Treatise.* Thereafter, though it was no longer considered as a research project, GSA support was generous and took a variety of forms.

Summary

Throughout its history, but especially since it began to aid graduate students directly, the research grants program has been one of the Society's most rewarding activities. In modern times, large sums of research money are commonly available from the National Science Foundation and many other sources. Such funds usually go only to institutions, however, and they are only for major projects and involve large overhead payments to the institutions, reimbursements which are forbidden with GSA grants. With a few other organizations, such as the Sigma Xi fraternity and the American Philosophical Society, GSA has consistently filled a real gap by aiding small short-term projects to be done by individuals. Many a graduate student has received his degree and been able to enter the geologic profession mainly because of GSA support for his required research. Many geologists who have since reached the pinnacles of professional success were helped at some point in their careers by GSA research support. The list is long, and it even includes two Nobel prize winners—Harold C. Urey and Willard F. Libby.

15

Libraries and Depositories[7]

Through the years, the Society management has had responsibilities for three distinct libraries. One of these, maintenance of a complete set of GSA's own publications, has been a continuing duty since the *Bulletin* first appeared. One complete set of bound volumes, plus a nearly complete backup set, is on display and available for in-place reference use at headquarters. A second library, consisting of the personal library of R.A.F. Penrose, Jr., came to GSA as part of his bequest. It is described in more detail below. The best of the collection is preserved in the Penrose Room and the Conference Room at headquarters. The remainder was placed with several geologic libraries on permanent loan, or is in storage at headquarters. The third collection, called here "Depository Library," surely rates near the top in the list of recurrent, persistent problems.

Penrose Library

R.A.F. Penrose, Jr., was a bibliophile. He read extensively, in his own field and in many others, but he also evidently loved rare books for their own sake. Some of the finest in his collection are uncut, obviously unread but treasured for their rarity, their titles, or their bindings. His main interests, aside from geology, seem to have been in early explorations, less of the western United States than of the rest of the world. He picked up many volumes during his travels to out-of-the-way places, but perhaps bought more from a favorite dealer in London than anywhere else.

At his death, Penrose's entire library became the property of the Society when the American Philosophical Society relinquished claim to its half-share of his office furniture, books, and other personalia. The library, consisting of several thousand volumes, lined most of the walls in the two New York headquarters buildings. It was housed in handsome cases, built to Penrose's specifications, of Uruguayan mahogany. The entire collection has been catalogued and appraised, and most of the valuable items have been repaired, rebound, or otherwise refurbished by an expert bookbinder. The library is maintained primarily as a memorial to the Society's benefactor, but it is available for study by accredited historians.

Representative of the volumes in the collection are the following:

Voyages and Travels in all parts of the World, John Pinkerton (1812, 17 volumes); *Oriental and Western Siberia*, T. W. Atkinson (1858); *Travels in Tartary, Thibet and China during the Years 1844-5-6*, M. Huc; *Expedition to Borneo of HMS* Dido *for the Suppression of Piracy*, H. Keppel (1846); *History of Java*, T. S. Raffles (1830); *Travels in the Interior of Brazil*, John Mawe (1812); *Travels to Discover the Source of the Nile*, James Bruce (1798); *A Voyage around the World, but More Particularly to the North-West Coast of America*, George Dixon (1789); *Voyages from Montreal*, Sir Alexander Mackenzie (1802). Volumes of exceptional historical value include *Essay on the Theory of the Earth*, Baron G. Cuvier (1827); *Report of the Exploring Expedition to the Rocky Mountains in the Year 1842*, John C. Fremont; *Principles of Geology*, Sir Charles Lyell (1842); *Modern Geography*, John Pinkerton (1804, 2 volumes); *The Mineral Kingdom*, Reinhard Brauns (1912); *Narrative of the Surveying Voyages of His Majesty's Ships* Adventure *and* Beagle, *between*

---

7. Taken in part from Fairchild (1932, p. 194–199) and in part from a 14-page file memorandum (by H. R. Aldrich?) titled "History of development of depository (exchange) list," May 1942.

the years 1826 and 1836, Describing their Examination of the Southern Shores of South America, the Beagle's Circumnavigation of the Globe, Captain P. Parker King, Captain Robert Fitz-Roy, Charles Darwin (1839).

Depository Library and the Exchange Program

Acquisition and maintenance of a research library was not one of the original objects of the Society, but dissemination of geologic knowledge through publication was. Accomplishment of this objective (publication) inevitably led to growth of a library.

The library was initiated by one of the founders, Professor C. H. Hitchcock, in May 1891, when the GSA was less than three years old. He wrote to all the Fellows asking for donations of their personal publications and photographs. This was the only effort ever made by the Society to gather published material from individuals; it resulted in numerous donations then and started a stream that has never completely ceased flowing.

Completion of Volume 1 of the *Bulletin* in May 1890 was a far more important event in growth of the library than was Hitchcock's request to the Fellowship. Donation of the *Bulletin* to 68 institutions began in 1891, and naturally brought donations in exchange. Establishment of an exchange system had been authorized at the second annual meeting in December 1889, but was not put into practice until the *Bulletin* had begun to appear regularly. The aim was to make GSA's published knowledge available to geologists throughout the world. Reciprocal exchanges of publications were of distinctly secondary importance and were not required. Nevertheless, most recipients of the *Bulletin* began to send their own publications in exchange. Even though not on the exchange list, numerous other publishers also sent their publications.

The number of exchanges maintained by the Society fluctuated widely through the years, depending perhaps more on the whims of Councils, committees, and secretaries than on anything else. Some changes, however, had rational reasons, such as attempts to balance the loss of income from subscription sales against the desire to disseminate knowledge, or communication difficulties during two world wars. At times, all domestic exchanges were eliminated; at other times, all foreign exchanges or those to specific continents were removed from the list.

Until 1935, exchanges and free subscriptions numbered less than 100. In 1935, and at least through 1941, the list was expanded to as many as 140 (and intermittent recommendations were made for several times that number). The expanded list, of course, reflects receipt of the Penrose fortune, which resulted in fewer financial worries and a greatly increased publication program. By 1955 the exchange list had grown to 348. In the following year, the Council acted on advice from the Policy and Administration Committee and voted to discontinue all domestic ex-

changes; foreign exchanges presented more political and other problems than the domestic ones so their discontinuance was delayed, but only for a short time when they too were stopped (Report of the Policy and Administration Committee, Council Minutes for April 1956).

The decision to end the exchange program was based almost entirely on economics. Fulfillment of exchanges and correspondence related thereto were costing the Society some thousands of dollars per year. More important, the free exchanges meant the loss of possible paid subscriptions. In a day when GSA publications had become required items in geologic libraries and when nearly all libraries could afford to pay, it seemed unnecessarily and uniquely generous on the part of the Society to give its publications away.

The influx of publications that followed closely on establishment of the exchange program must have alarmed the Council (and no doubt alarmed the Secretary, who had to accommodate the books within a small department of geology at Rochester). At the annual meeting in Columbus, December 1891, less than a year after the start of the program, Council named a committee to find a home for the accumulating literature. By the end of 1894 the Library Committee and the Secretary could report that a contract had been signed with the Case Library in Cleveland. The Case Library was an independent subscription library, established in 1880 as successor to the Cleveland Library Association, founded in 1848. It had no connection with Case School of Applied Sciences (later Case Institute of Technology), though both were funded by the same benefactor. The Case Library affiliated with Western Reserve University in 1924. By 1930 the merging of the Western Reserve and Case libraries had been completed, and most of the collections formerly owned by GSA were housed in the university's department of geology. Much later, in 1967, Western Reserve merged with Case Institute to become Case Western Reserve University.

The relatively elaborate contract placed all of the Society's library material on deposit with Case, title to remain with GSA. Case was to treat the material as it did its own collections, was to receive all additional exchange material, and was to catalogue, index, and bind all GSA property at its own expense. In 1897 the Society elected H. P. Cushing as its Librarian. He was succeeded in 1914 by F. R. Van Horn, who held office through 1917 (long after the GSA library had been sold!). Aside from memberships on the GSA Council, the main purpose in electing a Librarian was to provide someone to check periodically on the collection's welfare in Case Library hands.

The real reasons behind the choice of the Case Library as a repository are now unknown. The Society was probably wise in its decision to move the collection to the "West," because geologists concentrated in the eastern United States had ready access to many better and more complete collections. In addition to Case, Wisconsin and Chicago Universities had asked for the GSA collection and promised to care for it faithfully, though Chicago wanted it in perpetuity rather than as a loan. Cleveland has had its share of geologists but has never been known as a center of geologic research in the Midwest. Whatever the reasons, the Case Library in Cleveland was chosen as the recipient.

The entire library, by then consisting of 2,000 bound volumes and many unbound volumes, pamphlets, maps and charts, was sold outright to the Case Library in 1909. The purchase price was $2,250, plus an annual payment of $150 for future increments to the exchange material. By 1930 these additional payments, despite lapses and non-payments, had totaled $2,850.

Thus, by early 1910, the Society had disposed of all its interest in its library, beyond seeing to it that exchanges sent their material to the Case Library. Nevertheless, the Society's librarians, Professor Cushing and his successor at Western Reserve, J. E. Hyde, tried for many years to preserve the identity of the GSA collection and to see that it was adequately catalogued and cared for by both Case and Western Reserve librarians. Their devoted efforts met with only partial success, and the continuity and preservation of the collection gradually deteriorated.

Fairchild (1932, p. 199) makes much of the fact that the membership was never asked to vote on the sale of the library nor even informed of it. He concludes that the sale of Society property "was improper and probably illegal." It is indeed surprising to learn that the membership was asked to elect a Librarian each year until 1918, nine full years after the library was gone. The sale was not illegal, however, nor does it seem to have been improper. Each Council action that led up to it was recorded in the minutes. Even though they may not have been published then as they are now, the minutes were always available to any member who wished to study them. Moreover, all Council actions are ratified (i.e., made legal) every year by vote of the membership at annual meetings. Far larger business transactions than the sale of a $2,500 library are legalized regularly by this device. Purchases and sales of valuable real estate and of millions of dollars worth of securities through the years, for example, are among the actions by Councils, committees, or individuals that are seldom reported to or voted on by the membership until after the fact.

Reading between the lines of Fairchild's history, it seems reasonable to conclude that Fairchild was somehow unhappy with the performance of Edmund D. Hovey, his successor as the Society's Secretary, and that this may have led him to an intemperate attack on the legality of the library sale.

In 1944 the Society cancelled the contract with Western Reserve. Collection of the $150 annual payments from the university ceased, but the Society promised to offer surplus exchange items that were not needed by the U.S. Geological Survey after they had been used for compilation of the *Bibliography* (Council Minutes, 1944, p. 161). Neither GSA nor Western Reserve records are very clear as to the effects of this rather elusive promise, but it can be guessed that the inflow of material to the university library slowed abruptly. We do know that the Society stopped its exchange program completely in 1956 (Interim Proceedings for 1956, part II, p. 29). Moreover, though most of the items in the GSA library are probably still at Cleveland, the collection has been broken among several of the university's libraries and is no longer an identifiable entity (Letters of October 21, 1977, and May 26, 1978, from Samuel M. Savin, Chairman, Geology Department, to EBE).

Miscellaneous Book Problems

Despite the discontinuance of most exchange materials to Cleveland, many journals and books continued to accumulate at GSA headquarters. Early in 1973 and shortly after the Society had moved to its new headquarters, it was decided to dispose, once and for all, of the entire collection. Space was already at a premium, the journals were rarely consulted by the staff and never by the public, and they would be more useful and receive better care in established libraries than in the GSA offices. Accordingly, the journals were offered to geologic libraries by means of a full-page announcement in the newsletter, *The Geologist*. Terms were simple. Long runs of *American Journal of Science, Journal of Geology,* and *Economic Geology,* most of them morocco-bound, had come to GSA from the R.A.F. Penrose, Jr., personal library; these were to be treated as permanent loans. All other volumes were to be outright gifts.

Numerous librarians responded to the offer and asked for parts or all of the collection. Unfortunately for most of them, tragedy had struck the South Dakota School of Mines and the Pennsylvania Geological Survey during the summer of 1973, and the extensive and invaluable libraries of both institutions had been wiped out by floods. To help meet this greater need, virtually the entire GSA collection was divided between Rapid City, South Dakota, and Harrisburg, Pennsylvania. The few remaining items went to other requesters on a first-come, first-served basis, and apologies were sent to all others.

In addition to the journals and other serial publications that came by inheritance or through the exchange program, an almost continuous stream of books, to say nothing of authors' reprints, were sent to headquarters by publishers or authors. Though appreciated, their care and ultimate disposal, as well as provision of space and shelving, posed minor but constant problems for successive managements. So far as can be seen in fragmentary records, no firm policies seem to have ever been evolved. The books came, and as books will, they went. Some, perhaps most, were eventually given to nearby geologic libraries or to students; some undoubtedly entered the private libraries of GSA employees; still others were discarded or lost during headquarters moves.

In 1969 a determined effort to stem the flow of gift books was made; probably similar attempts had been made earlier. Most of the books came from their publishers with requests for review in the *Bulletin*. When such gifts were received, the donors were sent printed cards acknowledging the gift, telling them that the Society did not publish book reviews, and offering to return the gifts on request. This courtesy slowed the stream of books markedly but did not stem it completely. Shortly after the new journal *Geology* was established in 1973, its editor began to solicit books for possible review. This policy, not surprisingly, resulted in a renewed flow of books.

In recent years, most of the material still received from publishers or from former exchanges that refused to die has been sent to the editor of the *Bibliography of Geology* for preparation of bibliographic entries and use in the AGI-GSA GeoRef system. From there, the materials have gone to AGI's reference library or to the U.S. Geological Survey library.

GSA headquarters was never the place for a research library. There was neither internal nor external need for it. Even if there were a need, maintenance of an adequate library that would be used by the profession and would be a credit to the Society would require far more total resources, monetary and other, than would ever be available to any private society. It can be hoped that the problem of acquisition and disposition of journals and books will never recur, but the history to date suggests otherwise.

16

GSA: Good Friend and Neighbor

Internal Generosities

Like its principal benefactor, GSA has always been generous, yet reasonably discriminating both to its members and to its friends. In these days of self-supporting annual meetings and constantly rising costs for publications, younger members may find it hard to believe that their Society is generous. Many of them, however, have benefited directly from such gifts as research grants and free or cut-rate participation in Penrose Conferences and other activities. The fact is that for most of GSA's life all members have received more value in journals, book publications, and other perquisites than they have paid in dues and fees. In the earliest days the modest dues of $10 per year more than paid for the only publication, the quarterly *Bulletin*, for printing and mailing costs were low. This situation did not last long, however, and even before receipt of the Penrose fortune and expansion of the publication program, the Society produced the *Bulletin* at a net loss. Ever since the Penrose bequest became available, use of substantial proportions of its income in support of publications has been routine practice; all members have benefited from this policy through the years. Since about 1975 earnest and at times almost frantic efforts have been made to make the publication program truly self-supporting, but even today (1980) the members receive more publications for their money than do outside subscribers.

Relations with Other Organizations

Although it may at times have felt like the principal heir to a fortune who is besieged by distant relatives of the deceased, all seeking portions of the estate, GSA has maintained cordial relations with and been extraordinarily generous to other, less well-endowed organizations. Not only has it donated large amounts of money in support of its friends and neighbors but also it has welcomed them to its family gatherings and given them prominent space on its scientific programs; it also has donated thousands of copies of its publications and has aided other organizations in many other material ways. This generosity has not been one-sided by any means. Individuals and organizations have also given much to GSA. Their support is described in a separate section.

All gifts to its friends, in money or in kind, have furthered the Society's primary objective—advancement of the science. GSA is rightly proud of these outside contributions to geology, though it might privately wish for a little more public recognition of its generosity than is exhibited by some recipients. The total of all money gifts to other organizations, mainly since 1931, certainly exceeds a million dollars. More exact figures could be compiled by painstaking study of the records, but even these would be only approximate, for they would involve estimates as to the real values of enormous quantities of books and journals.

The largest single donation ever given by the Society was the $100,000 gift to finance the 1933 XVI International Geological Congress (Chapter 4). This was also the first donation after receipt of the Penrose bequest and was financed by income derived during settlement of the Penrose estate.

GSA supported publication programs of several societies, notably the Paleontological Society, the Mineralogical Society of America, and the Society of Economic Geologists, for many years. Prior to 1934 the contributions were made as a uniform percentage of each recipient's cost of publishing its periodical. In 1934 a more consistent and controllable plan was adopted. Thereafter, GSA agreed to pay 50% of publication costs for the Paleontological Society, the Mineralogical Society of America, and the Seismological Society of America to a maximum of $1,500 each per year. The limits were raised in later years, to as much as $10,000 at times, with different limits and percentages for recipients according to need. Regardless of details, the support program had a twofold purpose: (1) to help each society maintain its own publication and (2) to relieve GSA of part of the load of highly specialized papers.

In 1960 the Council voted to eliminate all subsidies to associated societies gradually by cutting the 1960 level of donations by 20% per year. The last donations were made in 1964, except for the American Geological Institute, which continued to receive for another decade sums that were somewhat larger than the per-capita "dues" expected from other constituent organizations.

Associated Societies

Since 1909 when the Paleontological Society was organized as a direct offshoot, GSA has welcomed other societies to join it as associates. The list has grown gradually, and even changed from time to time. In 1980 it included the following: Paleontological Society, Mineralogical Society of America, Society of Economic Geologists, Cushman Foundation, Geochemical Society, National Association of Geology Teachers, Geoscience Information Society, and Society of Vertebrate Paleontologists.

To this list must be added the Seismological Society of America, which has chosen for many years to retain its affiliation with GSA's Cordilleran Section and to meet with it annually rather than to associate with the parent Society. Like several other organizations, official associates of GSA or not, the Seismological Society received financial support for its journal from 1935 through 1964.

The term "associated society" is somewhat vaguely defined both in the Society's Bylaws and in practice. Association is accomplished on request and acceptance by the GSA Council. Implicit in the association is the agreement to take part in GSA annual meetings regularly or intermittently. Associated societies have a voice in the Joint Technical Program Committee. They plan and produce symposia or other segments of the program, and most of them hold their business and council meetings during the annual convocation. They pay no fees or dues, though their members share in the costs of meetings through the usual personal registration and other fees.

Other than the right to contribute to the technical programs just described, distinctions between associated societies and nonassociated ones are tenuous to nonexistent. Several dozen other groups—societies, institutes, national committees, even geology departments, meet regularly at the GSA annual meetings, and hold their business or social functions in facilities supplied in the overall meeting plan.

Paleontological Society

In December 1908 the formation of a Paleontological Society (PS) was proposed. It was to be a section of GSA, parallel with the Society's only geographic unit, the Cordilleran Section (Fairchild, 1932, p. 184). After the usual committee study, the GSA Council's acceptance, and necessary amendment to the constitution, the new society was launched just one year later and held its first annual meeting in conjunction with the proud parent. The paleontologists have met regularly with GSA ever since. From 1913 through 1954 the PS president automatically became Third Vice-President of GSA.

Because it lacked facilities to publish all of the worthy papers submitted to it in its own journal, the Paleontological Society was happy to have GSA publish a share of them in its *Bulletin*. (Nonpaleontologists would hold that on the basis of published evidence—Memoirs, the *Treatise, Bulletin* articles, and so on—the paleontological discipline has been exceedingly well treated through the years.) In partial reciprocation, GSA Fellows who were also elected to PS paid no dues to that organization. This custom, which was later paralleled by a similar one for the Mineralogical Society of America, was followed for many years, even after 1935 when GSA began to support its sister societys' publications on a regular basis. Evidently intended as tokens of appreciation for GSA's generosity, these agreements might well have been criticized (but never were) as prolonging the need for subsidies by reductions in PS and MSA dues incomes.

Mineralogical Society of America

The Mineralogical Society of America (MSA) was organized in 1919 and affiliated with GSA at the same time (Fairchild, 1932, p. 186). It already had its own publication, the *American Mineralogist,* started in 1916. Because GSA had no added publication expenses, except for MSA programs and abstracts for joint annual meetings, GSA agreed to pay the MSA treasury $3 for each MSA Fellow who was also a Fellow of GSA. Like the arrangement with PS, this custom lasted for many years, but when MSA was opened to two-member classes, the rules were changed, and only those who were Fellows (not members) of both organizations paid reduced dues directly to MSA.

Society of Economic Geologists

The Society of Economic Geologists (SEG) was organized and affiliated with GSA in December 1920, along with MSA. R.A.F. Penrose, Jr. was principal founder and first president (Fairchild, 1932, p. 186). SEG retained its affiliation and met with GSA for only three years. It then withdrew its formal affiliation as an associated society. It always remained a close friend of GSA, but thenceforward it tended to alternate its annual meetings between GSA and the American Association of Mining Engineers (AIME). In recent times, although it still meets with AIME on occasion, SEG holds most of its annual meetings with GSA and is again listed on the annual meeting programs as an associated society.

The Society of Economic Geologists received somewhat different treatment from the others. SEG never asked for or received help for its journal *Economic Geology,* which has from its inception been published by a closely related but separate organization, the Economic Geology Publishing Company. As mentioned in Chapter 3, the publishing company received a $25,000 bequest (less taxes) direct from Dr. Penrose, and GSA aided SEG in its successful claim for about $28,000 from the Penrose estate.

GSA played a large part in supporting SEG's *Annotated Bibliography of Economic Geology* in cash and in services through most of its life. This periodical was started in 1928; R.A.F. Penrose, Jr. was one of its founders. Professor Waldemar Lindgren, as chairman of SEG's Committee on the Annotated Bibliography, headed the project from its inception until his retirement in 1938. It was an SEG publication but was published through 1935 under the sponsorship and financial support of the National Research Council (NRC). In 1935 NRC's Committee on the Annotated Bibliography asked the GSA Council for a gift of $1,500 per year, in line with the support being given to other societies. The Council, feeling that the need for a specialized bibliography on ore deposits had lessened with GSA's broader *Bibliography and Index,* tabled the request (and NRC withdrew its own support of the *Annotated Bibliography*). Shortly thereafter, however, informal arrangements were made whereby the GSA bibliographic staff would produce annotated items for SEG's periodical as by-products of their regular work. This arrangement continued through 1967 when volume 38 (1965) was published as the last one of the series. Marie Siegrist, editor and chief compiler of the GSA series, contributed many of the SEG abstracts and was aided by numerous others, most of them volunteers. During that long period, GSA contributed heavily, not only by furnishing its bibliographic items but also in editing and printing of SEG's publication during parts of the period. Several other societies, as well as corporations and individuals, contributed heavily in cash. In his report to the 1962 Council, the Secretary estimated that

GSA's annual support to SEG in goods and services had been between $5,000 and $6,000. This was comparable to the average continuing support given to the other societies.

Society of Vertebrate Paleontologists

The Society of Vertebrate Paleontologists (SVP), established in 1941, received some monetary help for its own small journal for some years, but never on as large a scale or as regularly as the larger societies just described. The vertebrate paleontologists were greatly aided for many years, however, by GSA's support of the series of specialized bibliographies on their subject. This series, compiled by Professor C. L. Camp of the University of California, Berkeley, was supported in part by repeated research grants to Dr. Camp and in part by GSA publication of the resulting bibliographies in its Memoir series. In 1978 the compilation and publication were taken over by AGI as a by-product of the GeoRef bibliographic project. The SVP commonly meets with GSA only on alternate years. When it does, it is listed as an associated society and like the others takes part in formulating the technical program.

Since 1933 GSA has thus helped support its sister societies and their publications and devoted at least $500,000 to their direct support. Other scientific organizations, both in North America and abroad, have received modest financial support from GSA from time to time, but most was in the form of one-time gifts for specific purposes or needs and there is little point in listing them here.

American Geological Institute

GSA's support of the American Geological Institute (AGI) has been substantial ever since AGI was first conceived. As noted in numerous places in this book, support has taken many forms. Of these, direct financial support totaling at least $200,000 is the most visible. Significant as it was, it probably played a smaller part in AGI's growth and even its survival than the moral support given in the Institute's many times of stress.

The AGI, in reality a by-product of World War II, was conceived in 1943, but it did not become operational until long after the war, in November 1948. Earnest efforts by many groups, GSA among them, to increase the effectiveness of geology and geologists in a total war situation had some outstandingly good results (see section on War Efforts). More than anything else, however, the efforts had shown that geology was a house divided; there was no single group that could speak for the entire science and its workers, as the American Institute of Physics or the American Chemical Society, for example, could speak for their professions.

Initiative for the formation of a unifying organization was taken by the American Association of Petroleum

Geologists (AAPG), the largest geological society, but GSA became an early and staunch supporter. The AAPG called a meeting for April 10, 1943, inviting 100 geologists from the United States, representing most of the larger societies, to discuss the desirability of establishing a union of some kind.

AAPG's proposal was received with mixed feelings by most of the participants, who feared that their parent societies might lose their identities by submersion in a larger group. Nevertheless, enough progress was made to justify appointment of an organizing committee. By August 1943 the committee had prepared a draft constitution for a Geological Council of America, ready for presentation to the potential constituent societies. These, in turn, appointed study committees and began serious discussions. GSA itself was not entirely in favor of the union at the outset, largely because it needed assurance that affiliation would not endanger its tax-free status. After more thorough committee and Council study, however, GSA became one of the early and more enthusiastic supporters of the concept (Council Minutes for 1943, p. 264, and for 1945, p. 98).

The Institute was finally organized in November 1948 after a long and painful gestation period. The original need to provide for better use of geologists in war time had of course long since all but passed into history, but there were so many other needs that could be better served by a union than by any single group that there was never any fear that AGI would lack useful work to do.

In its formative years AGI was housed and sponsored by the National Academy of Sciences. Then, as now, its members were societies rather than individuals, but it was operated by and for those societies and all their members. Financial support came from the constituent societies and from individual and company donors. In later years, generous grants from the National Science Foundation for a variety of projects became by far the largest sources of funds. More recently, gradual withdrawal of most of the NSF support brought distress and painful reappraisal to an organization that had come to depend on grant money for its very existence.

The activities and accomplishments of AGI need not be recounted here. Some of them—the magazine *Geotimes,* the *Glossary of Geology,* the *Bibliography* and other products of GeoRef—are known to every geologist in the profession. Suffice it to say that AGI has survived and that it has grown to maturity, albeit slowly and painfully. It has not accomplished all the dreams of its founders, and may never do so. It has had its troubles, both internally and with its member organizations, some of which have been less than constant in their devotion. Despite all this, it has become a permanent and positive force in the geologic world; GSA can take pride in having played a strong part in that accomplishment. As a matter of fact, had not AGI been born or survived, the Geological Society of America as the largest organization purporting to represent all facets of geology would almost certainly have found itself shouldering many of the burdens that AGI has carried through the years.

Other Friends

As mentioned, numerous societies and other organizations, although not officially listed as associates, choose to hold their meetings at the same time and place as the GSA annual meeting, or just before or after it. All are welcomed. The list changes from year to year, but all participants appear in the annual meeting programs.

Aside from sharing its meetings, GSA helps provide guidance or advice to national and international societies, institutes, geological congresses, and other groups. For example, it sends its President or his delegate to International Geological Congress meetings, and plays much larger parts when the Congress meets in North America. It has a seat on AGI's governing board. Each annual *Membership Directory* contains a long list of GSA appointees to these and other groups. Some other appointments are not even listed in the *Directory.* These include a representative to the selection board for the Chrestien Mica Gondwanaland Medal of India and to a similar board for the annual James Furman Kemp Memorial Award to a Columbia University student (which representation is an unwritten duty of GSA's Executive Director).

Gifts of Books

In line with its major purpose of disseminating geologic information as widely as possible, the Society has been exceedingly generous with its publications throughout its life. By far the greatest contribution was the depository program described in Chapter 15. The program started with volume 1 of the *Bulletin* and has continued with regular gifts of all Society publications to scores of geologic libraries throughout the world for many years. Closely allied to this was the sale, at a nominal price, of the Society's library accumulated as a result of exchanges received from the depository program.

In addition to the depository program, abandoned in 1956, GSA has ever been proud and happy to hasten to the aid of libraries when they needed special help. For example, several libraries in Europe whose collections were destroyed in one or both of two world wars were sent complete sets of all GSA publications that were still in print. Similarly, earthquake-destroyed libraries in Japan were restocked in 1924; the war-destroyed library of the Geological Survey of Korea was replaced in 1956. More recently, fire- or flood-ravaged libraries in Brazil, South Dakota, and Pennsylvania have been replaced.

The Society has always been, and still is, a good neighbor to friends in need.

War Efforts

Of the five wars in which the United States has been involved since 1888, GSA played small but relatively significant parts in only the two World Wars. It was not involved in the short-lived Spanish-American War and very little in the Korean and Vietnam conflicts. In common with all individuals and organizations, however, it felt the effects of war-related changes in the economy and in social upheavals during the 1950s and 1960s. A few research grants were cancelled or postponed as recipients were called to service. Dues of enlisted personnel in the armed forces were remitted, although they continued to receive publications. Aside from these tokens, the Society's life style was unchanged during the wars in Asia.

World War I

Numerous geologists contributed significantly to the Allied cause during World War I—in the search for strategic minerals, in the application of geologic knowledge to trench warfare, in solution of water-supply problems, and in other ways both small and large.

The Society was one of the sponsors of the National Research Council (NRC), organized early in the war as an action arm of the National Academy of Science. In 1917 GSA contributed $400 to NRC toward financing a visit to France by Harry Fielding Reid. He was a geologist member of a multidiscipline committee sent to study the part science was playing in war. Reid's report on the mission, which was relatively unproductive, appears in the Council Minutes for 1917, p. 413. In 1918 the Society donated $2,000, a significant part of its total treasury, toward preparation of NRC reports on geology and geography for use by the Allied armies.

The Society conformed, of course, to regulations concerning conservation of paper and other materials. Having no paid staff, it had no problems with labor shortages. Several GSA members prepared books or pamphlets summarizing their personal knowledge of war-zone terrain or of possible uses of geology in war, but except for one or two *Bulletin* articles none of these were published by the Society.

World War II

The Society tried its best to aid the allies during World War II. On balance, it had reason to be proud of its contributions. A few of its efforts were relatively unsuccessful, in part because GSA was a small voluntary association of earth scientists who had no way of organizing, financing, or directing group efforts. It was also an exclusive group of mature geologists that was far from representing the majority of earth scientists in North America. Realization of this

fact and disappointment that GSA could not contribute in larger ways to the war effort led directly to democratization of the Society and establishment of a member class not long after the war's end. It also led eventually to establishment of the American Geological Institute as described earlier in this chapter.

The Society's desire to do its bit for the war effort was not entirely selfless. Many Fellows, including some of the older and wiser heads, felt that nothing the Society could do, geological or otherwise, could so much promote the science of geology in North America as to help make the war short and decisive. Others voiced fears as to the fate of the science itself or of the Penrose endowment for which GSA was responsible should the Allies lose the war. Still others wanted to maintain an active cadre of geologists ready to resume pure research that would have to be neglected during the war.

Committee on War Effort. As always, the Council turned to a committee to advise it and to guide the Society. The Committee on War Effort was appointed in December 1941, only a few days after Pearl Harbor; it remained active until war's end in 1945. It consisted of W. O. Hotchkiss, chairman, W. W. Rubey, and C. R. Longwell, three exceptionally clear thinkers who were each to contribute wisdom and energy to the conduct of the war on a far broader scale than the small part to be played by GSA. Later, the committee was further strengthened by addition of K. C. Heald and C. K. Leith, and Longwell became chairman. From its inception the Committee on War Effort concentrated on finding the most effective use of geology and of geologists in a great war. Some attention was also paid to the continuing production of new geologists, available for war–related service or for the postwar period. This objective could receive only secondary emphasis, however, because total war requires that a nation get along on what it already has in trained manpower and in basic knowledge.

Most of the committee's work lay in making contacts in the maze of agencies and bureaus and attempting to educate responsible officials as to the usefulness of geology and geologists in both civilian and military activities. It worked closely with similar groups in the AAPG, SEG, and other societies and disciplines, and with authorities in innumerable war agencies. The Secretary and headquarters staff contributed heavily to the committee's work. Many of the problems of best use of scientists and technicians were coordinated by the Office of Scientific Personnel (OSP), National Research Council. The committee, which had unusually wide latitude in making financial commitments for the Society, contributed a total of about $15,000 to support of the OSP during the war period.

Awareness of the values of geology was most certainly spread widely as a result of the educational efforts of GSA's committee and related groups. Tangible results beyond such awareness were mixed. Many of the geologists who were using other than geologic skills in the armed services were identified and a few were transferred to geologic activities, but many more were not. Efforts to protect younger geologists from the draft so that they could continue specialistic educations or could remain employed on war-related work as civilians were relatively successful. Success came a little late, however, and ended before the end of the war when virtually all able-bodied men under 26 became draft-eligible.

Other War-Related Activities. Aside from the continuing efforts of the Committee on War Effort to see that geology was used effectively, the Society contributed to World War II in many ways, and felt many of its effects, some good and some bad.

GSA contributed of its wealth directly by investing $650,000 in U.S. Government bonds, gladly accepting a significant reduction in interest income. Part of the loss in income was made up by a capital gain on the railroad and utility bonds that were sold to finance the purchase, but the motivation was entirely patriotic. It was probably the first time in its history that the Society had ever deliberately cut its income by switching investments.

As mentioned above, GSA also donated some thousands of dollars to continuing support of the Office of Scientific Manpower and like groups. Another effect on the Society's finances resulted from the near-cessation of research grants and consequent growth of a treasury surplus that was later available for inevitably increased postwar needs. A small handful of research projects that promised some kind of war-applicable results were financed, but nearly all potential research grantees were otherwise engaged for the duration.

The Society's first-ever cost-of-living pay increase went to the staff in 1941. Called a "bonus" of 5% on the first $3,000 of salary, it was supposed to be temporary, but actually became permanent, and was but the first of many cost-of-living salary increases that were to follow in the years to come.

The New York headquarters staff remained almost intact throughout the war, but the bibliographic staff in Washington nearly went out of business. This was to be expected, because the inflow of foreign literature had slowed to a trickle, and because the specialized knowledge of GSA's bibliographers was needed for uses more directly related to the war. Marie Siegrist, chief bibliographer, was paid for a small part of her time by GSA, but was totally immersed as a reference librarian and linguist for the Military Geology unit of the U.S. Geological Survey. Characteristically, she managed to keep up with entries for the GSA *Bibliography* in her "spare" time. Eleanor Tatge, assistant bibliographer, took full leave to work for the Beach

Erosion Board in the Department of Defense. Both women retained their re-employment and retirement rights in GSA, and both returned after the war.

Annual technical meetings for the years 1942 through 1944 were cancelled, for obvious reasons. The annual business meetings of the Society, required under the articles of incorporation, were held as usual, but at New York headquarters. Council meetings were continued but in abbreviated form and with more absentees than usual because so many councilors were engaged in war duties.

The publication program was affected in many ways. The *Bibliography*, shortened by lack of foreign literature and bibliographers, became a biennial. Offerings of new manuscripts for the *Bulletin* dwindled, restrictions were imposed on use of paper, and printing houses experienced labor shortages. The *Bulletin* and other publications were subject to censorship and none were sent to enemy-held countries. The exchange program with foreign publishers was largely suspended, as it had been during World War I. Rather than being saved for years for possible reuse, zinc cuts for halftones and line drawings went to the scrap pile as soon as used because their metal was needed for other purposes.

GSA cooperated with the American Association of Petroleum Geologists in helping distribute a series of special pamphlets. These included a bibliography of military geology, two booklets on the use of geologists in war, and one on the role of geology in World War I. Ernst Cloos of Johns Hopkins abstracted numerous papers on Germany's application of geology to its war machine. These were distributed widely by the Committee on War Effort to military officers and others.

Secretary Aldrich and the headquarters staff intensified their efforts in preparation of a directory of American geologists, a low-level, prewar project. Much progress was made despite problems caused by war-related address changes. The directory was never completed and published by the Society, but work on it continued past the end of the war in cooperation with the National Roster of Scientific and Specialized Personnel.

All in all, the Society's efforts to contribute to the war effort were profitable and praiseworthy. It emerged from the war with staff, treasury, and objectives intact. Few of the leaders or the membership realized that the old, relatively calm prewar days would never return. Like everyone else, the Society was entering a period of intensified and accelerated activity, of expansion, of economic, political, and social change, and above all a new recognition of the place of science in a changing world.

GSA: Recipient of Gifts from Its Friends

Generous as it has always been to its friends and neighbors, GSA itself has never lacked friends who gave generously of both time and money. The dollar value of all the time donated to GSA by hundreds of members—officers, councilors, committeemen, editors, manuscript reviewers, and others—is incalculable but it is enormous. Thirty years' free rental of Columbia University's building on West 117th Street, New York City, and of storage space in its library is also incalculable, but very large.

The U.S. Geological Survey has even been a staunch friend of the Society since its founding. It has shared in solving problems of production and costs of publishing some large and expensive maps; it provided library space for GSA bibliographers and shared the bibliographic coverage of the world literature of geology for almost half a century. It is the principal host for annual meetings that take place near any one of its major centers. It has lent members of its staff to act as GSA Editors for long periods, and has always been generous in allowing its geologists, from top officials downward, to serve the Society as officers or on committees. The USGS counterparts in Canada and Mexico have been equally generous; their contributions have been less only because more GSA population and activities center in the United States than in the neighboring countries.

Among institutions, the National Science Foundation has been the largest donor of money to the Society. In 1968 it granted $480,000 to alleviate a publication logjam. Because of its expert knowledge of the money needs in geologic research, it regularly provides a delegate to the Society's research grants committee. The significantly large and continuing contributions by the University of Kansas to support of the *Treatise on Invertebrate Paleontology* are described in the chapter on Society publications.

Gifts of money ranging from a few to thousands of dollars have been numerous, but their total addition to the Society's capital is miniscule as compared to the nearly $4,000,000 gift from R.A.F. Penrose, Jr. It is the Society's half of Penrose's entire fortune, augmented by growth and income from wise investment, that sets GSA apart from other learned societies that depend largely on membership dues for survival. As told in Chapter 10, another smaller but very sizeable personal fortune—that of Raymond C. Moore and his wife Lillian—will be added to the Society's wealth at some time in the future.

All gifts of money or securities or the many fine mineral specimens and other art objects given for the new building whether from living donors or from bequests have been gratefully acknowledged by the Council or its delegate, the Secretary (Executive Director). Those gifts whose uses were specified by the donors, such as endowments for the Penrose and Day medals, the Birdsall gift to the Division of Geohydrology, and the Harold T. Stearns gifts for research in the Circum-Pacific region, have been rigidly accounted for and their incomes applied to meet the donors' wishes. Unrestricted gifts have been applied to specific purposes by Council action or have simply been swallowed up by addition to the current operating fund.

In only one case was it impossible for the Society to follow the donor's wishes. In 1973 the August and Barbara Goldstein Foundation generously donated stock worth a little more than $10,000 asking that its income be used for decoration and furnishing of GSA's new home or for addition to the research grants program. The gift was kept intact, and before any of it could be used, the price of the donated stock plummeted close to zero as did the anticipated income from it.

In the late 1930s the Society received a small bequest left by Mrs. Ida Kunz Taggart, given in memory of her brother, George F. Kunz, the noted gem expert and an original Fellow of GSA. Mainly because the surviving family was in need, GSA relinquished its claim at the end of 1939 (Council Minutes for 1939, p. 259). To revive the analogy between R.A.F. Penrose, Jr., and the Society, this is almost certainly the action that Penrose would have taken in similar circumstances.

Nearly all significant gifts received by the Society have been announced to the membership. Some have been announced at appropriate gatherings such as the annual banquet or at the business meetings of the divisions, but most have been announced only in print, either in the annual reports of officers or in other news reports to the membership. Surprisingly, however, the Society maintains no running cumulative record of all the gifts it has received through the years—not in an engraved plaque and not even in the Society's accounting records. This omission is difficult to explain; to correct it now would also be difficult.

Although it has always welcomed unsolicited gifts and bequests, whether small or large, the Society has seldom sought gifts actively. The desirability of increasing the endowment funds has been discussed by the Council from time to time; the early decision to permit life commutation of dues by payment of a small lump sum on election to Fellowship was one of the few positive results of such discussions.

Since 1931, reluctance to seek more money was largely based on the fact that the Society felt comfortably rich with the Penrose endowment, whose income generally permitted it to do most of what it wanted to do on behalf of the science. No doubt its reluctance was also a matter of modesty and a feeling that begging was unseemly. Conversely, potential donors of large or small gifts have doubtless tended to eliminate GSA from their plans because they perceived it to be better off financially than many other worthy causes.

One serious attempt was made to seek additional funds. In 1958, toward the end of Henry Aldrich's regime as Secretary, a Committee on Growth and Support was appointed. Largely through the individual effort of its chairman, Professor Paul F. Kerr of Columbia, this committee worked diligently for several years planning and organizing a fund-raising campaign and drafting appeal brochures. The only concrete result was establishment in 1961 of the Henry R. Aldrich Publication Fund. Its object was to provide a revolving fund to be used in publication of book-length manuscripts that could not be accommodated in the regular program. Later the objective was changed, almost diametrically, to provide for reprinting of the few Memoirs and Special Papers that had proved to be best sellers. Appeals to the membership resulted in donations of $12,400, generous on the part of givers but producing an income too small to add significantly to the publication program. In 1962 the Council made the Aldrich Fund a viable one by contributing $50,000 to it. This action was in reality only a paper transfer to some of the Penrose endowment to another fund account. It did not increase the Society's wealth.

The plan continued, but haltingly, until 1973 when the fund was dissolved and its assets were credited to the Current Operating Fund. The concept was good, Aldrich received some of the honor he deserved, and a number of books were published that could not have been accepted otherwise. The fund was always difficult to administer, however, and it had become a bookkeeper's nightmare with too many decisions on allocations of publication costs and sales income based on arbitrary, spur-of-the-moment decisions. Even though the Aldrich Fund is gone, original contributors can be sure that their gifts were put to good purpose.

Alarmed by the effects of inflation and of rapidly rising costs, particularly of publications, the Council decided in November 1978 to make an all-out effort to seek a major increase in the Society's endowment base. A Fund Raising Task Force that included some of the outstanding financial minds in the Society was appointed. Later it became the Centennial Development Committee. By the end of 1980, the Geological Society of America Foundation was established by the Council as a legal entity. Dwight V. Roberts, an experienced and successful fund raiser, was employed by the Foundation to serve as its president. He began his duties March 2, 1981. Named to serve on the Board of Trustees were Caswell Silver, Chairman, Robert L. Fuchs, Michel T. Halbouty, John C. Maxwell, and Hollis D. Hedberg. John C. Frye was appointed Secretary-Treasurer. The Foundation is a separately incorporated organization with an initial goal of $9 million—$4 million for support of the various Centennial programs and $5 million to expand GSA's normal activities in advancement of the science (*GSA News & Information*, 1981, v. 3, no. 3, p. 37 and v. 3, no. 4, p. 49).

Part IV

A Brief Look Ahead

The future is obscure, but the Society moves confidently toward its second century of devotion to the ideals of its founders and of its principal benefactor, R.A.F. Penrose, Jr.

17

The Future of Geology and of the Society

Future of Geology

The science of geology has come a very long way since GSA was founded in 1888, almost a century ago. The body of knowledge has expanded astronomically. More important, the facts in that body of knowledge are not merely a random collection of bits and pieces; more than ever before they tend toward a coherent, oriented whole that is acceptable to geologists everywhere.

As Fairchild was completing his review of the Society's first four decades, he correctly foresaw a bright future for GSA, but his predictions as to the future of geology itself, viewed in retrospect, were somewhat clouded (Fairchild, 1932, p. 228). Fairchild believed that the two principal themes of geology in his generation—stratigraphy and paleontology—had already made their great discoveries and left only addition of details for future discovery. The thousands of stratigraphers and paleontologists who have made such great strides in their specialties since 1932 would rightly claim that they have added far more than mere detail to their chosen science.

Fairchild listed eight categories of great problems that were currently recognized and that awaited solution by scientists of the future. His categories were geologic time, the earth's future, the earth's interior, diastrophism, paleogeography, climatology, origin of life, and evolution.

With slight changes in nomenclature that have occurred since 1932, Fairchild's list indeed covers a large proportion of the scientific problems that have occupied geologists during the past half century. Much progress solving them, well known to all geologists and increasingly so

to the lay public as well, have been made. For example, far better estimates of the age of the earth and of the universe are available now than in 1932. Ages of rocks and fossils and of waters and woods can now be determined with relative accuracy, in contrast to the almost meaningless estimates of the past. Again, the facts of plate tectonics and of sea-floor spreading (combinations of two of Fairchild's categories, diastrophism and paleogeography) have revolutionized geologic thought and brought more comfort to more geologic minds than any single former discovery other than the story of evolution.

Though he rather clearly foresaw the directions future scientific inquiry would take, Fairchild's vision was perhaps blurred in two important respects. First, he believed strongly that all the problem categories he listed were in the province of scientists other than geologists—biologists, chemists, physicists, and mathematicians. By implication, he believed that most advances in the future would be made by those other scientists, not by geologists. He could not have foreseen the development of hybrid geologists, those trained and highly skilled in basic sciences other than geology. Nor could he have foreseen the trend toward collaboration between scientists with team attacks on research problems big and little, in contrast to the earlier methods of individual inquiry.

Second, Fairchild erred, perhaps, in failing to foresee the amazing growth in applied geology as compared to theoretical geology done for its own sake. This shortcoming, which is noticeable throughout his book, is partly a reflection of the times and partly of his own academic background. In more recent years, more and more of the world's total geologic effort is devoted to application of geologic facts and principles to practical ends—discovery and development of petroleum and mineral resources, war time uses, both strategic and tactical, prediction and prevention or avoidance of geologic hazards, and increasingly, solution of environmental factors. All these and more applications of geology are responsible for much of the growth in the profession of geology. They are also responsible for innumerable additions to the growth of basic knowledge, for no geologic investigation, whatever its primary purpose, can help but add new knowledge.

The future directions of geology over the next 50 or 100 years are no more clear to this historian than they were to Fairchild in 1932. It is easy to predict that details will be added at ever-increasing rates to every branch of the science. What will come in theoretical geology that will match or surpass the impact of the great discoveries of the past? No one can say specifically, but great discoveries are sure to come.

We are just beginning to understand the oceans and the materials in and beneath them. The makeup and character of the earth's interior is still enigmatic in large part. The origin of life is understood only in a general way, and study of organic evolution through the immensely long Precambrian Era has barely started. Man has landed on the moon, and we are already learning much about many other celestial bodies. Questions as to whether the wealth and effort to be expended on space exploration might be more usefully devoted to study of our own earth are raised by many. Certain it is, however, that space exploration will continue, will contribute much to man's knowledge of his universe, and will yield many facts that will add to our understanding of our planet.

Discoveries of specific new geologic truths comparable in scope to those of the past may be unpredictable now. On the other hand, practical applications of geology to the common good are sure to burgeon and their general nature can be predicted with some certainty.

Most conventional sources of fuel, except possibly coal, will be greatly reduced long before the end of the next century. Duration of conventional sources will have been prolonged by intense applications of geologic skills, by invention of new exploratory techniques, and by massive revisions of economic concepts, but eventual exhaustion is inevitable. Lack of oil and gas and shortage even of coal will have gradually come to be accepted as a way of life rather than the short-term "energy crisis" that most people of the late 20th century like to think will disappear with time. Unless the civilization we now know is to end, however, satisfactory substitutes for conventional energy sources will have been discovered and put to use in time. Discoveries and developments will be made by application

of geologic, chemical, and physical principles that are already known or remain to be discovered.

Some fair and rational balances between those who would protect the environment at all costs and those who would provide for production of the materials and fuels that are essential to civilization's survival will be reached and in far less than a century's time. Most of the facts needed in striking balances will come from geologists working in industry, in academia, and at all levels of government. Geologists with a flair for practical politics as well as for geology will be at a premium.

Long before the end of another century, the causes of earthquakes will be understood. It will be possible to predict their occurrence in time and place, at least on land, but whether predictions will be acceptable to and heeded by the populace is doubtful. Methods for alleviating the disastrous effects of earthquakes or of relieving stresses enough to avoid surface ruptures effectively will be developed and in use in some places.

Geologic "accidents" other than earthquakes will continue to occur so long as the earth's dynamic forces continue, but more of them will be predicted in advance. People will not change appreciably—they will still build on unstable slopes or on active faults or near volcanoes; they will continue to crowd rivers and shorelines. But geologists and engineers working together will have warned them of the hazards and will have taught them to design and build structures that will alleviate the effects of "accidents" when they occur.

The ignorance and consequent fear that now surround and inhibit the entire concept of nuclear energy will gradually be dispelled and nucleur reactors, probably of designs undreamed of now, will have taken their place in supplanting substantial parts of the world's power now supplied by conventional sources. Development of nuclear energy is largely in the realms of the physicists, chemists—and politicians—but geologists will have played large and essential parts in finding raw fissionable materials and in solving the equally vexing problem of safe waste disposal for millenia. Vast resources of geothermal energy, now largely untapped, will have been identified by geologists, and many will have been developed.

Hidden ore deposits will be discovered by methods not now envisioned, or known but considered uneconomic. Collaboration between biologists, chemists, and geologists will teach us far more about the relations between geology and human health. Chemical substances useful to man, though some of those uses not yet known, will be produced, but not all from the rare, highly concentrated deposits we still depend on today. Instead, they will come from dilute low-grade to trace dispersions in rocks or waters that are only of scientific interest now. Economic geologists, more aware than any others that theirs are nonrenewable, wasting assets, will not only intensify their efforts to find new sources of conventional raw materials but also will lead the way in identifying and finding substitutes and in promoting reuse of all metals and other materials.

Growth in the geologic population, at least as great as the overall population increase in North America, seems certain. Future geologists will be far better trained than those of the past or present, but largely trained in narrow specialties rather than with the breadth that characterized so many of the greats of former years. Intellectual giants there will be, but most of them will be far less visible than formerly, mainly because most of their accomplishments will be well known only to their own small groups of specialists.

Whether the present trend toward emphasis on the laboratory and away from the field is reversible is anyone's guess. Most certainly, however, the field work that is done will be highly sophisticated, with means of transportation and observation now unknown and with now undreamed of instruments that will be rivalled or surpassed only by those in future laboratories.

Future of the Society

In his preface to Fairchild's history of the Society's birth and adolescence, and dealing particularly with the opportunities and responsibilities inherent in the magnificent bequest from R.A.F. Penrose, Jr., Joseph Stanley-Brown wrote in 1931:

The connected story of the Society's successes in what might be called the first stage of its career, we now have in permanent form. It is certain that we stand on the threshold of a new era. It does not seem probable that what Longfellow in his well known poem to Agassiz calls "the rhyme of the universe," and the geologists call geology, will not furnish ample opportunity for research during the next half century. What this new era may bring forth no one may now know, but with such a background of training and achievement on the part of its members, supplemented by ample financial strength, it may be expected confidently that the next historian will be able to show that the new responsibilities were fully met and that The Geological Society of America came to be a still greater vitalizing force and a more powerful factor in the advancement of geologic science. [Fairchild, 1932, p. ix.]

Now, half a century later, it can be said with pride that the Geological Society of America has met with the responsibilities imposed by the Penrose gift and lived up to the ideals and dreams of the founders almost a century ago. It not only has preserved its share of the Penrose fortune but also has seen it grow, perhaps a bit too modestly and slowly to suit everyone, but healthily and safely. Far more important, the Penrose endowment has generated income of more than $12 million, three times the size of the original gift. It has distributed all this and much more toward advancement of the science. The Society's accomplishments—

in publishing research results, in providing forums for exchange of ideas at its annual meetings and Penrose Conferences, and in financial and moral support of sister organizations—are recounted at length in this book and need no summarization here.

The course the Society will take over the next 50 or 100 years is at least as difficult to predict as is the course of geology itself. The health and even longevity of the Society, its internal character, and above all the ways it will choose to advance the science will all be influenced by technical, social, and economic factors that cannot be foreseen clearly in 1980.

The following brief but broad statements, perhaps presented more positively than is justifiable, represent the considered opinions of but one historian and are not much more than educated guesses. Others, some with more knowledge of the Society and better foresight, will doubtless have other opinions that differs from these.

With increase in the geologic population, GSA will almost certainly continue to grow in numbers. It will also grow in wealth and in expenditures, at least enough to keep pace with inflation, and very possibly much faster by reason of continued wise investments and future additions to its endowment.

Present-day governance and management techniques will probably continue virtually unchanged. The Society will continue to be conservative, with slow evolutionary changes to meet changing needs of its members and of the world around it. It will never become an activist group, politically or technically.

GSA's future as an important publisher of geologic literature is least predictable of all its activities. The Society's publication program, summarized in length in Chapter 10 has occupied the lion's share of the Society's attention and resources since the very beginning. On it depends much of GSA's worldwide prestige and longevity and many of the advances in geologic knowledge. Much more of the Society's resources have consistently gone into its publications, directly or indirectly, than into any other of its activities. From the organizational meeting in 1888, every Council has devoted much of its best thought to considering the health and welfare of the publication program, as reported to it by the editors, publication staffs, and committees. This situation can be confidently expected to continue indefinitely.

For many reasons both internal and external the publication program is certain to change markedly over the years ahead. The fact that change will take place is predictable; the character of the change is not. Methods for communication of research results, of which printed publication is but one, are changing rapidly. More and more geologists are specialized in their interests, which means that most research results need be communicated to relatively few colleagues, even though the geologic population as a whole is increasing in numbers, along with massive increases in the total body of knowledge.

Reading and writing habits and methods are changing too. A world without books, or hard-copy journals, or libraries without shelves is unthinkable to many, yet publishing technology and economics are changing so rapidly that the possibility must be at least considered. The trend to microform publication and to computer-activated data-retrieval systems, distasteful as they are to some, seems inevitable.

How the Society or any other geologic publishing house will choose to cope with changes and advances in scientific communication methods is impossible to say, but solution of the publication problem will require the best efforts of many successive Councils and their advisers in the years to come.

Someone may ultimately devise a format for large meetings that permits more efficient use of time than obtains now. If this should occur, and be accepted, the whole concept of meetings would change. So long as the Society continues to try to offer all things to all people, however, multipurpose annual meetings will probably continue much as they are now, and for as long as GSA survives. They will doubtless also continue to grow in size until attendance reaches some optimum figure—10,000. Concurrent sessions will continue to multiply and individual sessions will be restricted to ever tighter specializations. Meetings will lengthen from three or four days to something like a full week. Field trips as essential parts of annual meetings will dwindle in number and may eventually be abandoned, as the number of cities that can accommodate a GSA convention narrows and as higher proportions of geologic research are done in the laboratory rather than in the field.

The single annual meeting may be split into several distinct meetings, either in sequence or separated in time and space. Section meetings will possibly also increase in size and numbers. They will present much more of the material that is now given at the main Society convention.

Other, more specialized societies will continue to multiply and to grow in membership. Some or many of these will associate with GSA. As each matures, it will seek more independence of action than is now available to associated societies, though not to the extent of separation from GSA's annual meeting or from any financial benefits that affiliation may entail.

Other activities of the Society, such as financial support of research projects and sponsorship of Penrose Conferences, will doubtless continue as they have in the past. Outside influences, such as greatly increased public support of basic research, at some time in the future might possibly reduce the need for private research funds, but it seems much more probable that Society support will need to increase, financed by GSA's own endowment or by contributions from its friends.

Recurrent Problems

Each Council routinely faces innumerable recurrent duties and problems. Most of these are solved easily and relatively quickly. Such duties as preparation of the budget, selection of committees, nominations for officers, bestowal of honors and awards, and review of reports from divisions, sections, and other groups are annual affairs. They all receive careful and sometimes prolonged consideration, but their solutions are commonly apparent.

Many other recurrent problems, however, seem never to be completely resolved. Time and again throughout the Society's history the same problems surface. They are discussed by the Council, special committees are appointed to study them in depth, they are again discussed by the Council, often at great length. Typically, they then submerge with little or no positive action, only to re-emerge a year or a decade later. Random examples of such problems include decisions on multi- versus single-class membership, criteria for selection of officers and councilors—and whether or not multiple slates of nominees should be offered—search methods for top-level employees, and the kind and degree of support to be provided to sister societies.

In addition to such basically administrative problems, several philosophical ones have persisted for decades. These include proposals for annual review articles, for memorial volumes (festschrifts), and the desirability of publicizing the Society or of popularizing geology. Perhaps most persistent and significant of all is the question as to whether the Society should play a more active role in guiding the directions of fundamental research.

These examples of unsolved, recurrent problems are logically included in a chapter on the future of the Society if only because each of them can be firmly predicted to occupy some of the time and energy of future governing bodies. More important, however, their very persistence through the years shows that they reflect chronic sore spots in the Society's anatomy; as such they require attention. Firm decisions as to the kind and degree of treatment for any of them are sure to have far-reaching effects on GSA's future character and well-being.

By far the greatest problem to be faced by GSA in the future is that of survival of a generalistic society in a specialistic world. Will it survive with somewhat the same form and purpose as we know it now? Will it wither in favor of more specialized groups? Will it subdivide with more divisions, or by spawning new societies? Will it perhaps displace the American Geological Institute with some super-union of American or even or international geological societies? Answers to none of these questions are apparent today, but this is not the last time they will be asked, from within and without the Society. On the answers will depend the future nature of the Geological Society of America.

18

Epilogue

The first 90-odd years of the Society's life described in these pages happens to have been almost equally divided into two epochs with markedly different characteristics. The first epoch extended from the founding in 1888 to 1931. It was marked by gradual growth in maturity, numbers, and prestige, in parallel with significant additions to the body of geologic knowledge.

The second epoch extended from 1932 into 1980. Initiated by GSA's inheritance of half of the R.A.F. Penrose Jr., fortune, virtually every activity in the ensuing years was affected by the relative freedom that resulted from the endowment's income. The second or Penrose epoch coincided in time with massive increases in knowledge, in new theories and new technologies, and in the geologic population.

The year 1980 appears to mark the beginning of a third epoch in Society history. How long it will last and what its characteristics will be must be left to the next historian, but it is reasonably safe to say that he will find evidence of enough changes in the Geological Society of America—anatomy, philosophy, activities, and products—to justify his description of a third great epoch.

Changes begun in 1980 will almost certainly result in major changes in the Society's major continuing activity—its publication program. As envisioned by the Council in early 1981, the principal publication products will continue in outward appearance much as they had evolved during the Penrose epoch. One notable exception will be the absence of the *Bibliography and Index of Geology,* which spanned most of the previous epoch but which will continue publication under the aegis of the American Geological Institute. The *Bulletin* and *Geology* seem likely to

continue much as they were before the near-demise of the *Bulletin* during the disastrous microfiche experiment of 1979–1980. Their greatly increased circulations, thanks largely to captive audiences of the entire GSA membership, should strengthen the financial positions of both journals; far more important, the larger circulations should add measurably to their impact on the science and to the prestige of their contributors.

The other elements of the publication program—books (including the *Treatise*), maps and charts, *Abstracts with Programs,* and others—should continue, on the whole, much as they have in the past. It is to be hoped, however, that improved business and accounting methods together with new techniques in editing, printing, and production will serve to arrest or at least ameliorate the ever-increasing sales prices of most GSA publications—price increases that have demonstrably cut into circulation and hence diminished the Society's place in the highly competitive science publication business.

While the general character of GSA's publications seems likely to continue much as it had evolved prior to 1980, the internal details of publications management and production will change in many important ways as a result of decisions taken in 1980 and 1981. Instead of a single Science Editor on the paid staff, submitted manuscripts will be judged and controlled throughout the publication process by one of a group of volunteer, absentee Science Editors. Responsibility for editing, proofreading, and style will rest largely with the appropriate Science Editor, with the authors, and with outside editors, rather than with the cohesive unit of in-house staff editors and production assistants that operated throughout the Penrose epoch. Changes to new (to GSA) printers are imminent; even warehousing and distribution will be in new hands.

The facts of change in the publication program are self-evident. The effects of those changes—in appearance, scientific content, public acceptance, and economic viability—are far from clear at this writing. Most of the changes will doubtless prove to have been good; they will surely be continued. Those changes that prove to have been mistakes will just as surely be abandoned by future Councils. Future historians may well fail to perceive immediate outstanding changes in the general character of the Society's publication series that relates to the 1980 beginning of the third epoch. On closer examination of the series, however, they can hardly fail to see marked changes in the program, both quantitative and qualitative, that began about 1980 and that continued into some now unknown but finite time in the future.

Many changes in the Society other than those in publications began to take form in 1980. Their success or failure will have as much to do with characterizing the third epoch as will the publications changes. Appointment of a new Executive Director is imminent. His personality and character, and those of his inevitable successors, especially in interactions with the Executive Committee, the Council, and the staff will have enormous impact on the Society's directions in the years to come.

The Society's plans to celebrate its 100th anniversary in 1988, centered around the Decade of North American Geology (D-NAG), took firm form in 1980. Possibly the single grandest scheme ever conceived by a scientific society, it will, if it comes up to its founders' dreams, set the stage for a quantum leap forward in the earth sciences. The D-NAG products and those that are generated by it in the future will alone establish 1980 as the beginning of a third epoch in the history of the Society. Success or failure of the centennial plans will depend in large part on the availability of large quantities of new money, the sources of which are not yet apparent. Success or failure will also depend on the whole-hearted cooperation of hundreds of authors and of their institutions, working over a long period of time, and

on enlightened management of a highly complex publication effort. This effort is to be superimposed on the Society's normal publication program, which has itself not been without management problems in the past.

Finally, introduction and continuation of a third epoch in Society history will require infusion of large amounts of new money. During the first epoch, GSA managed to be self-supporting and to do most of the things it wanted to do on a small income from membership dues and a few subscriptions to its journal. The Penrose endowment marked the beginning of the second epoch, and the constant flow of income from that endowment permitted all of the expansion in activities and products that gave unique character to the epoch and set it off from the earlier period. For most of the years since Penrose's death in 1931, his endowment provided the largest single source of money available to the Society.

This ability to depend on income from the Penrose endowment for a large part of GSA's activities no longer holds. Tax laws limit the proportion of total income that a society can generate from endowments, and inflation continues to eat away at the real value of any income, whether it be from dues, subscriptions, or other sources. The Penrose endowment is still well-managed, as it has always been, but is simply cannot keep pace with the Society's need for money to continue its established activities or to start new ones.

In an effort to solve the financial problems that existed at the time and to provide for even greater ones in the future, the Council in 1980 established the Geological Society of America Foundation. The first serious effort to seek gifts of money actively, the aim of the Foundation is to add very substantially to the overall endowment, first so that the Society can pay for the Decade of North American Geology and its products, and second to permit GSA to exert even greater impact on the course of the science than it has ever done in the past.

The search for new funds is thus a key element in the Society's third epoch. Its degree of success and the uses to which the hoped-for funds are put will determine the features that make the epoch unique in Society history.

First Constitution and By-laws of The Gelogical Society of America.

As adopted December 27, 1889 (*Bulletin*, v. 1, p. 571)

PREAMBLE.

The Fellows of The Geological Society of America, organized under the provisions of the Constitution approved at Cleveland, Ohio, August 15, 1888, and adopted at Ithaca, New York, December 27, 1888, hereby ordain the following revised Constitution:

ARTICLE I.—NAME.

This Society shall be known as THE GEOLOGICAL SOCIETY OF AMERICA.

ARTICLE II.—OBJECT.

The object of this Society shall be the promotion of the Science of Geology in North America.

ARTICLE II.—MEMBERSHIP.

SECTION 1. The Society shall be composed of Fellows, Correspondents, and Patrons.

SEC. 2. Fellows shall be persons who are engaged in geological work or in teaching geology, and resident in North America.

Fellows admitted without election, under the PROVISIONAL CONSTITUTION, shall be designated as ORIGINAL FELLOWS on all lists or catalogues of the Society.

SEC. 3. Correspondents shall be persons distinguished for their attainments in geological science, and not resident in North America.

SEC. 4. Patrons shall be persons who have bestowed important favors upon the Society.

SEC. 5. Fellows alone shall be entitled to vote or hold office in the Society.

ARTICLE IV.—OFFICERS.

SEC. 1. The *Officers* of the Society shall consist of a President, First and Second Vice-Presidents, a Secretary, a Treasurer, and six Councilors.

These officers shall constitute an Executive Committee, which shall be called the *Council.*

SEC. 2. The *President* shall discharge the usual duties of a presiding officer at all meetings of the Society and of the Council. He shall take coguizance of the acts of the Society and of its officers, and cause the provisions of the Constitution and By-Laws to be faithfully carried into effect.

SEC. 3. The *First Vice-President* shall assume the duties of President in case of the absence or disability of the latter. The *Second Vice-President* shall assume the duties of President in case of the absence or disability of both the President and First Vice-President.

SEC. 4. The *Secretary* shall keep the records of the proceedings of the Society, and a complete list of the Fellows, with the dates of

their election and disconnection with the Society. He shall also be the Secretary of the Council.

The *Secretary* shall co-operate with the President in attention to the ordinary affairs of the Society. He shall attend to the preparation, printing, and mailing of circulars, blanks, and notifications of elections and meetings. He shall superintend other printing ordered by the Society or by the President, and shall have charge of its distribution under the direction of the Council.

The *Secretary*, unless other provision be made, shall also act as *Editor* of the publications of the Society, and as *Librarian* and *Custodian* of the property.

SEC. 5. The *Treasurer* shall have the custody of all funds of the Society. He shall keep an account of receipts and disbursements in detail, and this shall be audited as hereinafter provided.

SEC. 6. The Society may elect an *Editor*, to supervise all matters connected with the publication of the transactions of the Society under the direction of the Council, and to perform the duties of Librarian until such time as, in the opinion of the Council, the Society should make that an independent office.

SEC. 7. The *Council* is clothed with executive authority, and with the legislative powers of the Society in the intervals between its meetings; but no extraordinary act of the Council shall remain in force beyond the next following stated meeting, without ratification by the Society. The Council shall have control of the publications of the Society, under provisions of the By-Laws and of resolutions from time to time adopted. They shall receive nominations for Fellows, and on approval by them shall submit such nominations to the Society for action. They shall have power to fill vacancies *ad interim* in any of the offices of the Society.

SEC. 8. *Terms of Office.*—The President and Vice-Presidents shall be elected annually, and shall not be eligible to re-election more than once until after an interval of three years after retiring from office.

The Secretary and Editor shall be eligible to re-election without limitation.

The term of office of the Councilors shall be three years; and these officers shall be so grouped that two shall be elected and two retire each year. Councilors retired shall not be re-eligible till after the expiration of a year.

ARTICLE V.—VOTING AND ELECTIONS.

SEC. 1. All *elections* shall be by ballot. To elect a Fellow, Correspondent, or Patron, or to impose any special tax shall require the assent of nine-tenths of all Fellows voting.

SEC. 2. Voting by *letter* may be allowed.

SEC. 3. *Election of Fellows.*—Nominations for fellowship may be made by two Fellows, according to a form to be provided by the Council. One of these Fellows must be personally acquainted with the nominee and his qualifications for membership. The Council will submit the nominations received by them, if approved, to a vote of the Society in the manner provided in the

By-Laws. The result may be announced at any stated meeting; after which notices shall be sent out to Fellows elect.

SEC. 4. *Election of Officers.*—Nominations for office shall be made by the Council. The nominatons shall be submitted to a vote of the Society in the same manner as nominations for fellowship. The results shall be announced at the Annual Meeting; and the officers thus elected shall enter upon duty at the adjournment of the meeting.

ARTICLE VI.—MEETINGS.

SEC. 1. The Society shall hold at least two stated meetings a year—*Summer Meeting*, at the same locality and during the same week as the annual meeting of the American Association for the Advancement of Science, and a *Winter Meeting*. The date and place of the Winter Meeting shall be fixed by the Council, and announced by circular each year within a month after the adjournment of the Summer Meeting. The programme of each Meeting shall be determined by the Council, and announced beforehand, in its general features. The details of the daily sessions shall also be arranged by the Council.

SEC. 2. The Winter Meeting shall be regarded as the *Annual Meeting*. At this, elections of Officers shall be declared, and the officers elect shall enter upon duty at the adjournment of the Meeting.

SEC. 3. *Special Meetings* may be called by the Council, and must be called upon the written request of twenty Fellows.

SEC. 4. *Stated Meetings of the Council* shall be held coincidently with the Stated Meetings of the Society. Special meetings may be called by the President at such times as he may deem necessary.

SEC. 5. *Quorum.*—At meetings of the Society a majority of those registered in attendance shall constitute a quorum. Five shall constitute a quorum of the Council.

ARTICLE VII.—PUBILCATION.

The Serial publications of the Society shall be under the immediate control of the Council.

ARTICLE VIII.—AMENDMENTS.

SEC. 1. This Constitution may be amended at any annual meeting by a three-fourths vote of all the Fellows, provided that the proposed amendment shall have been submitted in print to all Fellows at least three months previous to the meeting.

SEC. 2. By-Laws may be made or amended by a majority vote of the Fellows present and voting at any annual meeting, provided that printed notice of the proposed amendment or By-Law shall have been given to all Fellows at least three months before the meeting.

BY-LAWS.

CHAPTER I.—OF MEMBERSHIP.

SEC. 1. No person shall be accepted as a Fellow unless he pay his initiation fee, and the dues for the year, within three months after ratification of his election. The initiation fee shall be ten (10) dollars and the annual dues ten (10) dollars, the latter payable on or before the annual meetings, in advance; but a single prepayment of one hundred (100) dollars shall be accepted as commutation for life.

SEC. 2. The sums paid in commutation of dues shall be invested and the interest used for ordinary purposes of the Society during the payer's life, but after his death the sum shall be covered into the Publication Fund.

SEC. 3. An arrearage in payment of annual dues shall deprive a Fellow of the privilege of taking part in the management of the Society, and of receiving the publications of the Society. An arrearage continuing over two (2) years shall be construed as notification of withdrawal.

SEC. 4. Any person eligible under Article III of the Constitution may be elected Patron upon the payment of one thousand (1,000) dollars to the Publication Fund of the Society.

CHAPTER II.—OF OFFICIALS.

SEC. 1. The *President* shall countersign, if he approves, all duly authorized accounts and orders drawn on the Treasurer for the disbursement of money.

SEC. 2. The *Secretary*, until otherwise ordered by the Society, shall perform the duties of Editor, Librarian, and Custodian of the property of the Society.

SEC. 3. The Society may elect an *Assistant Secretary.*

SEC. 4. The *Treasurer* shall give bonds, with two good sureties approved by the Council, in the sum of FIVE THOUSAND dollars, the faithful and honest performance of his duties, and the safe-keeping of the funds of the Society. He may deposit the funds in bank at his discretion, but shall not invest them without authority of the Council. His accounts shall be balanced as on the thirtieth day of November of each year.

SEC. 5. In the selection of Councilors the various sections of North America shall be represented as far as practicable.

SEC. 6. The minutes of the proceedings of the Council shall be subject to call by the Society.

CHAPTER III.—OF ELECTION OF MEMBERS.

SEC. 1. Nominations for fellowship may be proposed at any time on blanks to be supplied by the Secretary.

SEC. 2. The *form* for the nomination of Fellows shall be as follows:

In accordance with his disire, we respectfully nominate for Fellow of the Geological Society of America:

(Full name)

(Address)

(Occupation)

(Branch of Geology now engaged in, work already done, and publications made)

(Degrees, if any)

(Signed by at least two fellows)

The form when fill is to be transmitted to the Secretary.

SEC. 3. The Secretary shall bring all nominations before the Council, at either the Winter or Summer Meeting of the Society, and the Council shall signify its approval or disapproval of each.

SEC. 4. At least a month before one of the stated meetings of the Society, the Secretary shall mail a printed list of all approved nominees to each Fellow, accompanied by such information as may be necessary for intelligent voting. But an informal list of the candidates shall be sent to each fellow at least two weeks prior to distribution of the ballots.

SEC. 5. The Fellows receiving the list will signify their approval or disapproval of each nominee, and return the lists to the Secretary.

SEC. 6. At the next stated meeting of the Council the Secretary shall present the lists, and the Council shall canvass the returns.

SEC. 7. The Council, by unanimous vote of the members in attendance, may still exercise the power of rejection of any nominee whom new information shows to be unsuitable for fellowship.

SEC. 8. At the next stated meeting of the Society the Council shall declare the results.

SEC. 9. Correspondents and Partons shall be nominated by the Council, and shall be elected in the same manner as Fellows.

CHAPTER IV.—OF ELECTION OF OFFICERS.

SEC. 1. The Council shall designate three candidates for each office.

SEC. 2. The *form* for the nomiation and election of officers, unless otherwise provided by the Council, shall be as follows:

The Council nominates for Officers of the Geological Society of America, for the ensuing year, the following persons:
(The voter will indicate his preference out of each of the sets of names below by erasing the two other names in each set, or will substitute the name of his choice.)

For President,	1. 2. 3.
For First Vice-President,	1. 2. 3.
For Second Vice-President,	1. 2. 3.
For Secretary,	1. 2. 3.
For Treasurer,	1. 2. 3.
For Councilor,	1. 2. 3.
For Councilor,	1. 2. 3.

The Secretary shall mail a copy of this ballot to each Fellow, who after making up the list will return it to the Secretary.

SEC. 3. At the winter meeting of the Council, the Secretary shall bring the returns of ballots before the Council for canvass, and during the winter meeting of the Society the Council shall declare the results.

SEC. 4. In case a majority of all the ballots shall not have been cast for any candidate for any office, the Society shall by ballot at such winter meeting proceed to make an election for such office from the two candidates having the highest number of votes.

CHAPTER V.—OF FINANCIAL METHODS.

SEC. 1. No pecuniary obligation shall be contracted without express sanction of the Society or the Council. But it is to be understood that all ordinary, incidental and running expenses have the permanent sanction of the Society, without special action.

SEC. 2. The creditor of the Society must present to the Treasurer a fully *itemized* bill, *certified* by the official ordering it, and *approved* by the President. The Treasurer shall then pay the amount out of any funds not otherwise appropriated, and the receipted bill shall be held as his voucher.

SEC. 3. at each annual meeting, the President shall call upon the Society to choose two Fellows, not members of the Council, to whom shall be referred the books of the Treasurer, duly posted and balanced to the close of November thirtieth, as specified in the By-Laws, Chapter II, Section 4. The *Auditors* shall examine the accounts and vouchers of the Treasurer, and any member or members of the Council may be present during the examination. The report of the Auditors shall be rendered to the Society before the adjournment of the meeting, and the Society shall take appropriate action.

CHAPTER VI.—OF PUBLICATIONS.

SEC. 1. The Publications are in charge of the Council and under its control.

SEC. 2. One copy of each publication shall be sent to each Fellow, Correspondent, and Patron, and each author shall receive thirty (30) copies of his memoir.

CHAPTER VII.—OF THE PUBLICATION FUND.

SEC. 1. The Publication Fund shall consist of moneys paid by the general public for publications of the Society, of donations made in aid of publication, and of the sums paid in commutation of dues, according to the By-Laws, Chapter I, Section 2.

SEC. 2. Donors to this fund, not Fellows of the Society, in the sum of two hundred dollars, shall be entitled without charge, to the publications subsequently appearing.

CHAPTER VIII.—OF ORDER OF BUSINESS.

SEC. 1. The order of Business at *Annual Meetings* shall be as follows:

   (1) Call to order by the Presiding Officer.
   (2) Introductory ceremonies.
   (3) Statements by the President.
   (4) Report of the Council.
   (5) Report of the Treasurer, and appointment of the Auditing Committee.
   (6) Declaration of the results of the ballot for officers of the next ensuing Administration.
   (7) Declaration of the results of the ballot for new Fellows.
   (8) Announcement of the hour and place for the Address of the last ex-President.
   (9) Necrological notices.
   (10) Miscellaneous announcements.
   (11) Business motions and resolutions and disposal thereof.
   (12) Reports of committees and disposal thereof.
   (13) Miscellaneous motions and resolutions.
   (14) Presentation of memoirs.

SEC. 2. At an *adjourned session*, the order shall be resumed at the place reached on the prevous adjournment, but new announcements, motions, and resolutions will be in order before the resumption of the business pending at the adjournment of the last preceding session.

SEC. 3. At the *Summer Meeting* the items of business under numbers (4), (5), (6), (8), (9) shall be omitted.

SEC. 4. At any *Special Meeting* the Order of Business shall be (1), (2), (3), (7), (10), followed by the special business for which the meeting was called.

Present Constitution and Bylaws
of
The Geological Society of America, Inc.

CONSTITUTION

Article I
NAME AND SEAL

The legal name of the Society is "The Geological Society of America, Inc."; it is known as "The Geological Society of America." The corporate seal of the Society is circular in form and bears the name of the Society and the founding date of 1888.

Article II
PURPOSE

The purpose of the Society is the promotion of the science of geology by the issuance of scholarly publications, the holding of meetings, the provision of assistance to research, and other appropriate means. The Society also cooperates with other bodies having similar objectives and assists more recently formed societies interested in the specialized branches of geology.

Article III
MEMBERSHIP

Membership in the Society consists of Honorary Fellows, Fellows, and Members, all of whom have full and equal voting rights.

Article IV
MANAGEMENT

The affairs of the Society shall be managed by officers and councilors duly elected at regular intervals from the voting membership of the Society. As provided in the Certificate of Incorporation, as amended, the number of councilors shall be not less than ten (10) nor more than twenty-four (24), as may be provided from time to time by the Bylaws.

Article V
ANNUAL CORPORATE MEETING

The annual corporate meeting of the Society for the election of officers and councilors and for such other business as may properly come before the meeting shall be held at such time and place as the Council may from time to time prescribe.

Article VI
STATUTORY OFFICE

The statutory office of the Society shall be in the City, County, and State of New York, as required by the Certificate of Incorporation.

Article VII
TAX-EXEMPT STATUS

The affairs of the Society shall at all times be managed in such a way as to preserve and safeguard its tax-exempt status.

Article VIII
BYLAWS

Bylaws not inconsistent with this Constitution or with the Certificate of Incorporation shall be adopted at the time of the adoption of this constitution and may be amended as therein provided.

Article IX
AMENDMENTS

Amendments to this Constitution proposed not less than sixty (60) days before a lawfully held annual corporate or special meeting of the Society by a majority vote of the Councilors present at a lawfully held meeting of the Council, or by a petition signed by one hundred (100) of the membership of the Society, shall be set forth in the notice of the annual corporate or special meeting and may be adopted by a majority of the membership present in person or by proxy at that meeting.

BYLAWS

Article I
MEMBERSHIP

1. **Members**

Candidates for Member shall be persons who have either a Bachelor's degree with a major in geology or a related science, or equivalent training through practical experience, *or* have an active connection with geology through employment in geological work, or in teaching geology, or through status as a graduate student in geology. By delegation of Council authority to the Committee, Members shall be elected by the Committee on Membership upon production of satisfactory evidence, such elections to be ratified by the Council at its next regular meeting following the Committee action.

2. **Student Associates**

Candidates for Student Associates may be any full-time undergraduate or graduate student enrolled in a degree-granting institution and majoring in geology or related sciences, on satisfactory recommendation. Student Associates may not vote in Society elections. Graduate students may, if they wish, be enrolled as full Members with all voting and other privileges.

3. **Fellows**

Fellows shall be elected by the Council from the Members of the Society.

4. **Honorary Fellows**

Honorary Fellows shall be chosen from among distinguished geologists. Save in exceptional circumstances, the candidates shall be residents outside North America.

5. Applications
Applications for membership shall be transmitted in writing to the executive director on forms provided by the Society.

6. Resignations and Terminations
Resignations and terminations from membership or Student Associateship shall be transmitted in writing to the executive director. A Fellow or Member who has resigned in good standing may re-apply for membership by writing directly to the Committee on Membership. Such members will retain their former status and original date of membership, but the years when not active will be excluded from the total years of membership when viewing eligibility for exemption (Article VI, Section 4).

7. Suspensions
Any membership or Student Associateship in the Society may be suspended or terminated by the Council, after a hearing or opportunity to be heard, for conduct deemed prejudicial to the interests of the Society. A Fellow, Member, or Student Associate whose dues are in arrears for one (1) fiscal year will be automatically suspended.

Article II
MANAGEMENT OF THE SOCIETY
1. Management
The management of the affairs and the property of the Society shall be the responsibility of the Council which, pursuant to the Certificate of Incorporation, as amended, shall consist of not less than ten (10), nor more than twenty-four (24) councilors, as may be provided from time to time by these bylaws. The number of councilors is hereby set at sixteen (16) who shall consist of three (3) elected officers and thirteen (13) other voting members, one of whom shall be the immediate past president.

2. Officers
The elected officers of the Society shall be a president, a vice-president, and a treasurer. The appointed officers shall be an executive director and such others as the Council may from time to time determine, each to hold office at the pleasure of the Council or for such period of time as the Council shall designate. As required by law, the salaries of officers shall be fixed by the affirmative vote of a majority of the entire Council.

3. Election of Officers
The president, the vice-president, and the treasurer shall be elected at each annual corporate meeting of the Society, to hold office until the next annual corporate meeting and until their respective successors shall have been elected and qualified. They shall at the same time also be elected to the Council as provided in Section 4.

4. Election of Council
At each annual corporate meeting of the Society, the voting member who shall be elected president for one (1) year shall also be elected to the Council for two (2) years, to hold each such office until his successor shall have been elected and qualified; the voting members who shall be elected vice-president and treasurer shall also be elected to the Council, to hold office until the next annual corporate meeting and until their respective successors shall have been elected and qualified; four (4) other voting members shall be elected to the Council, to hold office for three (3) years and until their respective successors shall have

been elected and qualified. Each councilor shall be at least twenty-one (21) years of age.

5. Re-elections
The president and the vice-president shall not be eligible for re-election to their respective offices until at least three (3) years have elapsed from the expiration of their terms of office. The treasurer shall be eligible for re-election without limitation.

6. Vacancies
Any vacancy occurring in any of the elective offices of the Society shall be filled by the Council until the next annual corporate meeting.

7. Meetings of Council
The Council shall hold at least two (2) regular meetings during each calendar year for conducting the business of the Society, and such other special meetings as the Council shall prescribe. A majority of the members of the Council shall constitute a quorum, but less than a quorum shall have power to adjourn any meeting. Any such adjourned meeting may be reconvened without further notice.

8. Appointed officers
The Council shall appoint an executive director, who shall serve as secretary of the Society, and such other officers as the Council may from time to time determine.

9. Executive Committee
The Council may appoint from among its members an Executive Committee which shall be empowered to act for the Council between its regular meetings. All actions of the Executive Committee shall be reported to the Council for its information, and, except as to action on any item of business which the Council shall have theretofore specifically authorized the Executive Committee to act, shall be subject to ratification by the Council at its first subsequent meeting. The Executive Committee shall consist of at least three (3) but not more than five (5) members of the Council and shall include the president, the vice-president, the Budget Committee chairman, and normally the immediate past president. The president shall be chairman of the Executive Committee and one-half or more of its members shall constitute a quorum.

10. Expenses of the Council
Members of the Council may be reimbursed from the funds of the Society for their traveling expenses when attending meetings of the Council, but only the treasurer may be paid honoraria for his special services.

11. Annual Report
At each annual corporate meeting of the Society, the president shall submit to the membership on behalf of the Council a report upon the affairs of the Society for the year just closing, and this shall include a report verified by the president and treasurer or by a majority of the Council setting forth all information required by law with respect to the last preceding fiscal year. There shall be submitted to each annual corporate meeting a resolution providing for the ratification and confirmation of all acts of the Council during the preceding year.

12. Indemnification and Insurance Therefor
Every person shall be indemnified by the Society against reason-

able xpenses, including attorney's fees, actually and necessarily incurred by him in connection with the defense of any action, suit, or proceeding in which he is made a party by reason of the fact that he, his testator, or intestate is or was an employee, other than an officer, of the Society, or in connection with any appeal therein, except in relation to matters as to which he shall be adjudged in such action, suit, or proceeding to be liable for negligence or misconduct in the performance of his duties as such employee. Such right of indemnification shall not be deemed exclusive of any other rights to which such employee may be entitled apart from this Article. Councilors and officers, and former councilors and officers, of the Society shall be indemnified in the manner and to the extent permitted by law, and the Society shall purchase and maintain insurance for their indemnification as permitted by law, when, as, and if the Council shall so direct.

Article III
NOMINATION AND ELECTION OF OFFICERS AND COUNCILORS

1. Nominations

At least one hundred and twenty (120) days before each annual corporate meeting of the Society, the Council shall prepare and cause to be mailed to the voting membership of the Society a list of nominations chosen from the voting membership for the officers and councilors to be elected for the ensuing year, which names shall normally be chosen from those submitted to the Council by the Committee on Nominations. Any one hundred (100) of the voting membership of the Society may submit in writing to the executive director not less than sixty (60) days before the annual corporate meeting alternative nominations for one (1) or more of the positions to be filled. A list of all nominations shall be included in the notice of the annual corporate meeting. In addition to this method, space shall also be provided on the official ballot for write-in votes.

2. Official Ballot

An official ballot containing the nominations of the Council, special nominations received by the executive director in accordance with Section 1, and space for write-in votes, shall be given personally or mailed to the voting membership of the Society with the notice of each annual corporate meeting. If the notice is given personally or by first-class mail, it shall be given not less than ten (10) nor more than fifty (50) days before the date of the meeting; if mailed by any other class of mail, it shall be given not less than thirty (30) nor more than sixty (60) days before such date.

3. Elections

The officers of the Society and new members of the Council shall be elected at each annual corporate meeting in accordance with the procedure outlined in Article VII.

Article IV
POWERS AND DUTIES OF OFFICERS

1. The President

The president shall preside at all meetings of the Society, of the Council, and of the Executive Committee. He shall represent the Society on all appropriate occasions. He shall be *ex officio* a member of all committees of the Council.

2. The Vice-President

The vice-president shall have and assume the powers and duties of the president in the event of the absence or disability of the president.

3. The Treasurer

The treasurer, under the direction of the Council, shall collect and disburse all funds of the Society except those for which other provision shall have been made in the bylaws and in rules or resolutions of the Council. All funds, securities, and other investments of the Society shall be deposited in the name of the Society in the custody of a bank or trust company designated by the Council. The treasurer shall keep records of all receipts and disbursements and other financial transactions and of the funds, securities, and other investments of the Society. The treasurer shall submit an annual report to the Council as directed by the laws of the State of New York.

4. The Executive Director

The executive director shall attend meetings of the Council and act as the secretary-to-the-Council and keep the records of its proceedings; shall attend meetings of the Executive Committee when requested; shall be responsible to the Council for the work of the permanent staff of the Society; shall have the power and perform the duties of secretary of the Society; shall have custody of the corporate seal of the Society and shall affix and attest it as directed by the Council; and shall perform such other duties as the Council may direct.

5. Execution of Instruments

All agreements, conveyances, transfers, obligations, certificates, and other instruments and documents may be executed and delivered or accepted on behalf of the Society by the president, the vice-president, or by any other person thereunto authorized by Council; provided, however, that all written contracts (other than operational contracts, or other contracts involving moneys or obligations of $1,000 or less) must first be submitted to legal counsel and the executive committee for approval. The corporate seal fo the Society may be affixed to any such document or instrument, and attested, by the executive director.

6. Assistant Executive Director

If and so long as the Society shall have an assistant executive director, he shall have and may exercise all of the powers delegated to him by the executive director.

Article V
COMMITTEES OF COUNCIL

1. Committees of Council

The Council shall appoint standing committees as listed in Section 2; and the president, on the advice of the Council, may appoint special committees as may from time to time be deemed necessary. All such committees shall be advisory in character and shall report to and act under the direction of the Council. The Council may reimburse the members of such committees for the traveling expenses which they incur when attending regular meetings of their committees authorized by the Council. The president, or a representative designated by him, shall be an *ex officio* member of every committee.

2. Standing Committees

The Council shall appoint each year at its fall meeting the following standing committees:

Audit	Investments
Budget	Membership
Committee on Committees	Nominations
Geology and Public Policy	Penrose Conferences
Treatise on Invertebrate	Program Review
Paleontology Advisory	Publications
Headquarters Advisory	Research Grants
Honors and Awards	

In the case of the Committee on Committees, whose members serve a term of one year, the newly elected vice-president shall suggest to the Council the names of the persons to be appointed to that Committee.

In the case of the Membership Committee, appointments shall be made from the Fellowship of the Society.

3. Duties

The duties of each standing committee and of each special committee shall be determined by the Council. Each standing and special committee shall make its report directly to the Council at such times as the Council may direct.

Article VI
FINANCES

1. Fiscal Year

The fiscal year of the Society shall coincide with the calendar year.

2. Annual Dues

Dues shall be payable on the first day of January each year. The various options involving different groupings of publications and the cost of options each year shall be as determined by the Council.

To retain membership in the Society, all Fellows, Members, and Student Associates must pay the stated dues.

3. Exemption from Dues

The following categories of Members and Fellows shall be exempt from dues:

Honorary Fellows;

Penrose and Day medal recipients;

Life members (no longer used, this category includes members who remitted dues for life with a lump-sum payment);

Fellows and Members who have reached the age of 70 years and have paid annual dues for 30 years;

Fellows and Members who have reached the age of 65 years and have paid dues for 25 years, and who specifically request exemption from payment of further dues; and

Disabled members, as determined by the Committee on Membership on a case-by-case basis, for the duration of their disability.

Each Fellow or Member in any of these exempt categories shall be entitled to all the privileges of membership.

4. Arrears

Fellows, Members, and Student Associates whose dues have not been paid during the fiscal year to which they pertain shall be deemed in arrears, and they shall not be entitled to the Society rights and privileges until such dues have been paid. On receipt of full remittance, the member will be sent all back issues of Society publications to which he would have been entitled had he not fallen in arrears. Fellows, Members, or Student Associates whose dues have not been paid for one (1) fiscal year shall be suspended from membership in the Society.

A Fellow, Member, or Student Associate who has been suspended for nonpayment of dues (Article I, Section 7) may reapply for membership by writing directly to the Committee on Membership. Such members, if reinstated, will retain their former status and original date of membership, but the years when not active will be excluded from the total years of membership when viewing eligibility for exemption.

5. Penrose Bequest

The Penrose Bequest, established by the will of the late Richard A. F. Penrose, Jr., "shall be considered an endowment fund the income of which only to be used and the capital to be properly invested." A reserve fund, entitled "Penrose Bequest Reserve Fund," which shall consist of amounts accumulated from time to time, as directed by the Council, out of the income of the Penrose Bequest, has been established to offset possible losses in investments or for such other purposes as the Council may determine. No part of the Reserve Fund held for any purpose shall exceed such amount as shall be permissible under the laws of the State of New York and with respect to the Society's right to tax exemption.

6. Bequests

The Society may accept gifts and bequests at the discretion of the Council.

7. Audit

At each annual corporate meeting a firm of certified public accountants shall be appointed by the voting members present in person or lawfully represented at such meeting to audit the financial affairs of the Society. At the close of each fiscal year, the auditors shall audit and examine the records, accounts, vouchers, and financial transactions of the treasurer and the other officers and custodians of the Society, and the funds, securities, and other investments of the Society in the custody of the treasurer and other officers and custodians of the Society, and shall prepare a balance sheet and a statement of revenues, expenses, and changes in fund balances, prepared in accordance with generally accepted accounting principles, and shall report thereon to the Council at its first regular meeting of the new fiscal year. The report of the auditors shall be filed in the office of the executive director and shall be open to the inspection of the members at all times.

Article VII
MEETINGS

1. Annual Corporate Meeting

The annual corporate meeting of the Society shall normally be held in the month of November in each year at a location to be determined by the Council.

2. Special Meetings

Special meetings of the Society, except as otherwise required by law, may be called at any time by the president or a majority of the full Council. In any case, such meetings may be convened by voting members entitled to cast one hundred (100) votes or ten percent of the total number of votes entitled to be cast at such meeting, whichever is lesser, who may, in writing, demand the call of a special meeting specifying the date and month thereof, which shall not be less than two nor more than three months from the date of such written demand. The executive director upon receiving the written demand shall promptly give notice of such meeting, or if he fails to do so within five days thereafter, any voting member signing such demand may give such notice.

3. Notice of Meetings
Notice of the place, date, and hour of every meeting of the membership shall be given personally or by mail to each member entitled to vote at such meeting. If the notice is given personally or by first-class mail, it shall be given not less than ten (10) nor more than fifty (50) days before the date of the meeting; if mailed by any other class of mail, it shall be given not less than thirty (30) nor more than sixty (60) days before such date. Unless it is an annual corporate meeting, the notice shall indicate that it is being issued by or at the direction of the person or persons calling the meeting. Notice of a special meeting shall also state the purpose or purposes for which the meeting is called. If mailed, notice is given when deposited in the United States mail, with postage thereon prepaid, directed to the voting member at his address as it appears on the record of members, or, if he shall have filed with the executive director a written request that notices to him be mailed to some other address, then directed to him at such other address.

4. Quorum
At all meetings of the Society (except a special meeting for election of councilors called on the demand of the voting membership pursuant to the New York Not-For-Profit Corporation Law), members entitled to cast one hundred (100) votes or one-tenth (1/10th) of the total number of votes entitled to be cast, whichever is lesser, present in person or by proxy, shall constitute a quorum; but less than a quorum shall have power to adjourn any meeting. Any such adjourned meeting may be reconvened without further notice.

5. Tellers and Inspectors of Election
At any time prior to a meeting, except a special election required by the Not-For-Profit Corporation Law, the president may appoint two (2) or more tellers and inspectors of election to serve at such meeting. Vacancies below two (2) shall be filled by the presiding officer at the meeting. The tellers and inspectors of election at each meeting shall count the voting membership present in person or lawfully represented at such meeting and report the result orally to the meeting and shall also canvass the ballots cast on every vote taken by ballot and report the results in writing to the meeting.

6. Ballots
The Council shall cause to be prepared and mailed to the voting membership with the notice of each annual corporate meeting of the Society, the official ballot described in Article III, Section 2, of these bylaws. The official ballot shall also set forth other propositions that may come before the meeting. Ballots shall be returned by the voting membership to the executive director. Associated with the ballot shall be printed a proxy con-

taining language authorizing another person to attend the meeting and vote as directed in the ballot, this proxy to be signed by the voting member submitting the ballot. Such proxy shall also contain a provision that it will be accepted and exercised only if it is received at the headquarters office of the Society not later than ten (10) days before the date of the meeting.

7. Joint Meetings
The Council may at its discretion arrange for the holding of meetings of the Society jointly with other similar bodies for the discussion of matters of mutual interest, but attendance at the annual corporate meeting of the Society shall always be restricted to the membership of the Society.

Article VIII
REGIONAL SECTIONS
1. Regional Sections
Honorary Fellows, Fellows, Members, and Student Associates resident in specified geographic regions may, with the approval of the Council, organize as sections of the Society. The duly elected officers of such sections shall be responsible to the Council for the conduct of the affairs of their sections and shall render to the Council an annual report from their sections. Each section shall be known as "The . . . . . . Section of The Geological Society of America."

All members and Student Associates residing within the geographical limits of a section will normally be members of that section and will be so listed unless they choose membership in another section, as described in the following paragraph.

Those who reside within one section, but whose principal interests lie within the geographical limits of another section, may become members of that section, with all rights and privileges pertaining thereto, on written request to the Society. In so doing, they relinquish their rights and privileges in the section where they reside unless they desire to continue non-voting affiliation thereto.

Affiliation with additional sections for purposes of receiving meeting notices and attendance at meetings, but without privileges of voting or of holding office, is available upon written request to the Society.

Student Associates may not vote in section elections; they may serve as *conferees* on section committees if the section's bylaws so permit.

2. Bylaws
Each section may adopt for its own use bylaws approved by the Council which are consistent with the Certificate of Incorporation and the Constitution and Bylaws of the Society.

Insofar as is possible, the Council will ensure that the bylaws for the several sections are consistent with one another.

To become effective, proposed changes in a section's bylaws must receive management board approval, acceptance by the voting affiliates of the section, and ratification by Council.

3. Regional Meetings
Each section may, with the approval of the Council, organize

annual and special meetings within its own region. Circulation of notice of all such meetings shall be the responsiblity of the sections, and notice shall be sent to the membership within the respective region as well as to the headquarters of the Society and to all officers and members of the Council.

4. Finances
Each section may, with the approval of the Council, make its own arrangements for the raising of the necessary funds for the proper conduct of its operations. This may be done by means of the registration fees charged for attendance at meetings of the section. Account of all such funds shall be rendered to the Council at the end of each fiscal year.

Article IX
DIVISIONS OF THE SOCIETY
1. Divisions of the Society
Honorary Fellows, Fellows, Members, and Student Associates from particular branches of geology, and Honorary Fellows, Fellows, Members, and Student Associates concerned with the application of geology to other fields of endeavor may, with the approval of the Council, organize as divisions of the Society. The duly elected officers of such divisions shall be responsible to the Council for the conduct of the affairs of their divisions and shall render to the Council an annual report from their divisions. Each division shall be known as "The . . . . . Division of The Geological Society of America." Affiliation with any division or divisions is available to any member or Student Associate upon written request to headquarters.

2. Bylaws
Each division may adopt for its own use bylaws approved by the Council which are consistent with the Certificate of Incorporation and the Consitution and Bylaws of the Society. Insofar as is possible, the Council will ensure that the bylaws for the several divisions are consistent with one another.

To become effective, proposed changes in a division's bylaws must receive management board approval, acceptance by the voting affiliates of the division, and ratification by the Council.

3. Divisional Meetings
Each division may organize meetings as a part of and in association with the annual meetings of the Society. If a division desires to organize special meetings apart from the annual meetings of the Society, prior approval must be sought and obtained from the Council. Notice of meetings held at the time of the annual meetings of the Society shall be included in the general Society notice of such meetings. Notice of any special divisional meetings shall be the responsibility of the division itself which shall send such notice to its enrolled membership as well as to the headquarters of the Society and to all officers and members of the Council.

4. Finances
Each division may, with the approval of the Council, make its own arrangements for the raising of the necessary funds for the proper conduct of its operations. Account of all such funds shall be rendered to the Council at the end of each fiscal year.

Article X
PUBLICATIONS
Consistent with the stated purpose of the Society of promoting the science of geology, the Council shall arrange for the regular publication and distribution of scholarly papers as journal articles, memoirs, special volumes, or other media for the transmittal of information on the science of geology. The Committee on Publications shall be responsible to the Council for the high standards that are to be maintained in all such publications. Voluntary assistance with the work of producing the publications through such tasks as critical reviewing shall be a general responsibility of the membership.

Article XI
ASSOCIATED SOCIETIES
Any national or international society that has aims consistent with those of The Geological Society of America, that is, the advancement of the science of geology, may, with the approval of the Council, associate itself with the Society for the purpose of cooperation in annual, sectional, or divisional meetings, publications, and in other appropriate ways.

Such Associated Societies shall not have organic connection with the Society nor need their membership be confined to members and Student Associates of the Society. They may be known as "Associated Societies" in relation to The Geological Society of America.

Article XII
ALLIED ORGANIZATIONS
The Council may arrange for cooperation, not inconsistent with the Certificate of Incorporation and the Constitution and Bylaws of the Society, with other organizations having similar interests for mutual benefit and for the further promotion of the science of geology.

Article XIII
AMENDMENTS TO THE BYLAWS
Amendments to the bylaws may be made by a majority vote of the councilors present at any regular meeting of the Council. Every such amendment may be repealed by a majority of the voting membership present in person or lawfully represented at the next annual corporate or special meeting of the Society. Any amendment proposed not less than sixty (60) days before the next annual corporate or special meeting of the Society by a petition signed by one hundred (100) of the voting membership of the Society shall be set forth in the notice of such meeting and may be adopted by a majority of the voting membership present in person or lawfully represented.

Appendix C

Directory to
Proceedings, Abstracts, and Memorials of
The Geological Society of America

During the years since the Society's founding in 1888, Proceedings, Abstracts, and Memorials have appeared in various of the Society's publications. This directory should guide users to the right publications.

Proceedings: Proceedings of the Society cover business and other transactions (some of which may also appear in the Society's Annual Report). Memorials to deceased members have generally been published wherever and whenever the Proceedings have been published. Abstracts of papers presented at meetings have at times appeared wherever the Proceedings have appeared; at other times, Abstracts have been issued separately.

In recent years, Proceedings have included: annual reports of officers and committees; actions of the Council, the Society's governing body; reports on both business and scientific sessions of the annual meeting; reports on the presentation of medals and awards; standing statistics; proceedings of the Society's regional sections and its divisions; and proceedings of meetings of Section E (Geology and Geography) of the American Association for the Advancement of Science.

In earlier years, Proceedings also included the proceedings of other societies associated with The Geological Society of America, for example, the Society of Vertebrate Paleontology and The Paleontological Society.

Proceedings are of interest not only to members of the Society but to historians of science.

Memorials: A memorial is an essay on the life and career of a deceased member (usually a Fellow) of the Society. The Memorial ordinarily includes a complete bibliography of his works. Memorials are of interest to biographers, historians of science, and to scientists conducting research on topics on which the deceased has published.

Abstracts: Abstracts published by the Society are of all papers presented at technical meetings in which the Society participates. They include abstracts of papers presented by members of societies holding joint meetings with the Geological Society of America. Because of the wide range of subjects covered in the papers presented, the Abstracts are of interest to all geologists whether or not they are members of one of the participating societies. Increasingly through the years, published adstracts have come to be considered as citable items in the literature of geology.

LOCATING KEY

Year	Proceedings	Abstracts	Memorials
1889–1932	*Bulletin, v. 1-44*	*Bulletin*, v. 1-44	*Bulletin*, v. 1-44
1933-1937	*Proceedings,*	*Proceedings*	*Proceedings*
1938-1960	*Proceedings*	*Bulletin*, v. 49-71	*Proceedings*
1961	*Bulletin*, v. 72	Special Paper 68	None
1962	*Bulletin*, v. 73-74	Special Paper 73	*Bulletin*, v. 73-74
1963	*Proceedings*	Special Paper 76	*Proceedings*
1964	*Proceedings*	Special Paper 82	*Proceedings*
1965	*Proceedings*	Special Paper 87	*Proceedings*
1966	*Proceedings*	Special Paper 101	*Proceedings*
1967	*Proceedings*	Special Paper 115	*Proceedings*
1968	*Proceedings*	Special Paper 121	*Proceedings*
1969	*Annual Report of Officers and Committees**	*Abstracts with Programs*, v. 1	Memorials series
1970	*Annual Report of Officers and Committees**	*Abstracts with Programs*, v. 2	Memorials series
1971	*Annual Report of Officers and Committees**	*Abstracts with Programs*, v. 3	Memorials series
1972	*Annual Report of Officers and Committees**	*Abstracts with Programs*, v. 4	Memorials series
1973	*Annual Report for 1973**	*Abstracts with Programs*, v. 5	Memorials series
1974	*Geology* or *GSA News & Information*†	*Abstracts with Programs*, v. 6	Memorials series
1975	*Geology* or *GSA News & Information*†	*Abstracts with Programs*, v. 7	Memorials series
1976	*Geology* or *GSA News & Information*†	*Abstracts with Programs*, v. 8	Memorials series
1977	*Geology* or *GSA News & Information*†	*Abstracts with Programs*, v. 9	Memorials series
1978	*Geology* or *GSA News & Information*†	*Abstracts with Programs*, v. 10	Memorials series
1979	*Geology* or *GSA News & Information*†	*Abstracts with Programs*, v. 11	Memorials series
1980	*Geology* or *GSA News & Information*†	*Abstracts with Programs*, v. 12	Memorials series

*Published as a separate, complete publication.
†Published in segmented form where noted. Additional items collected and bound in annual report for year noted for distribution to libraries upon request.

INFORMATION FOR	CAN BE FOUND IN
Annual meetings and scientific sessions	Programs for the individual meetings
Annual corporate (business) meetings	Scripts for the individual meetings (Original copy of script bound in the minute book for the year)
Presentation of the Penrose and Day medals and the division awards	*Bulletin*
Standing statistics	GSA Yearbook or Membership Directory for the given year.
Memorials	Memorials series
Proceedings and scientific sessions for section and division meetings	Reports to Council from section and division secretaries, section annual meetings programs, and GSA annual meeting programs.

Past Presidents

1889 James Hall*
1811–1899

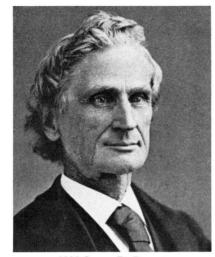

1890 James D. Dana*
1813–1895

1891 Alexander Winchell*
1824–1891

1892 and 1909 G. K. Gilbert*
1843–1918

1893 J. William Dawson
1820–1899

1894 T. C. Chamberlin*
1843–1928

1895 N. S. Shaler*
1841–1906

1896 Joseph Le Conte*
1823–1901

1897 Edward Orton*
1829–1899

* Original Fellow

Past Presidents

1898 J. J. Stevenson*
1841–1924

1899 B. K. Emerson*
1843–1932

1900 G. M. Dawson
1849–1901

1901 C. D. Walcott*
1850–1927

1902 N. H. Winchell*
1839–1914

1903 S. F. Emmons*
1841–1911

1904 J. C. Branner*
1850–1922

1905 Raphael Pumpelly*
1837–1923

1906 I. C. Russell*
1852–1906

Appendix D

Past Presidents

1907 C. R. Van Hise*
1857–1918

1908 Samuel Calvin*
1840–1917

1910 Arnold Hague
1840–1917

1911 William M. Davis*
1850–1934

1912 H. L. Fairchild*
1850–1943

1913 Eugene A. Smith*
1841–1927

1914 George F. Becker*
1847–1919

1915 Arthur P. Coleman
1852–1939

1916 John M. Clarke
1857–1925

Past Presidents

1917 Frank D. Adams
1859–1942

1918 Whitman Cross
1854–1949

1919 J. C. Merriam
1869–1945

1920 I. C. White*
1849–1927

1921 James F. Kemp*
1859–1926

1922 Charles Schuchert
1859–1942

1923 David White
1862–1935

1924 Waldemar Lindgren
1860–1939

1925 William B. Scott
1858–1947

Past Presidents

1926 Andrew C. Lawson
1861–1952

1927 Arthur Keith
1864–1944

1928 Bailey Willis
1857–1949

1929 Heinrich Ries
1871–1951

1930 R.A.F. Penrose, Jr.
1863–1931

1931 Alfred C. Lane
1863–1948

1932 Reginald A. Daly
1871–1957

1933 C. K. Leith
1875–1956

1934 W. H. Collins
1878–1937

Past Presidents

1935 Nevin M. Fenneman
1865–1945

1936 W. C. Mendenhall
1871–1957

1937 Charles Palache
1869–1954

1938 Arthur L. Day
1869–1960

1939 T. Wayland Vaughan
1870–1952

1940 Eliot Blackwelder
1880–1969

1941 Charles P. Berkey
1867–1955

1942 Douglas Johnson
1878–1944

1943 E. L. Bruce
1884–1949

Past Presidents

1944 Adolph Knopf
1882–1966

1945 Edward W. Berry
1875–1945

1946 Norman L. Bowen
1887–1956

1947 A. I. Loversen
1894–1965

1948 James Gilluly
1896–1890

1949 Chester R. Longwell
1887–1975

1950 William W. Rubey
1898–1974

1951 Chester Stock
1892–1950

1952 Thomas S. Lovering
1896–

Past Presidents

1953 Wendell P. Woodring
1891–

1954 Ernst Cloos
1898–1974

1955 Walter H. Bucher
1888–1965

1956 George S. Hume
1893–1965

1957 Richard J. Russell
1895–1971

1958 Raymond C. Moore
1892–1974

1959 Marland P. Billings
1902–

1960 Hollis D. Hedberg
1903–

1961 Thomas B. Nolan
1901–

Past Presidents

1962 M. King Hubbert
1903–

1963 Harry H. Hess
1906–1969

1964 Francis Birch
1903–

1965 Wilmot H. Bradley
1899–1979

1966 Robert F. Legget
1904–

1967 Konrad B. Krauskopf
1910–

1968 Ian Campbell
1899–1978

1969 Morgan J. Davis
1898–1979

1970 John Rodgers
1914–

Past Presidents

1971 Richard H. Jahns
1915–

1972 Luna B. Leopold
1915–

1973 John C. Maxwell
1914–

1974 Clarence R. Allen
1925–

1975 Julian R. Goldsmith
1918–

1976 Robert E. Folinsbee
1917–

1977 Charles L. Drake
1924–

1978 Peter T. Flawn
1924–

1979 Leon T. Silver
1925–

Past Presidents

1980 Laurence L. Sloss   1981 Howard R. Gould
1913–   1921–

REFERENCES

Albritton, C. C., editor, 1963, The fabric of geology: Stanford, California, Freeman, Cooper and Company, 374 p.

Albritton, C. C., Jr., editor, 1967, Uniformity and simplicity—a symposium on the principle of the uniformity of nature: Geological Society of America Special Paper 89.

American Philosophical Society, 1976, Year Book 1975. The annual report of the society. Repeats and updates much useful information on history, members, finances, and the like.

Arnold, L. B., 1977, A historical perspective, in Schwarzer, T. F., and Crawford, M. L., American women in geology (symposium): Geological Society of American Geology, v. 5, no. 8, p. 493–494.

Berkey, C. P., 1936, Unveiling of the bust of Richard Alexander Fullerton Penrose, Jr.: Geological Society of America Proceedings for 1935, p. 53–56.

——editor, 1941, Geology, 1888–1938: Fiftieth anniversary volume: Geological Society of America, 578 p.

Cannon, H. L., and Davidson, D. F., editors, 1967, Relation of geology and trace elements to nutrition—A symposium: Geological Society of America Special Paper 90, 68 p.

Chadwick, G. H., 1945, Memorial to Herman LeRoy Fairchild [1850–1943]: Geological Society of America Proceedings for 1944, p. 185.

Cross, Whitman, and Penrose, R.A.F., Jr., 1895, Geology and mining industries of the Cripple Creek district, Colorado: U.S. Geological Survey 16th Annual Report, pt. II, p. 1–209.

Engel, A. J., James, H. L., and Leonard, B. F., editors, 1962, Petrologic studies: A volume in honor of A. F. Buddington: Geological Society of America, 660 p.

Fairbanks, H. R., and Berkey, C. P. 1952, Life and letters of R.A.F. Penrose, Jr.: Geological Society of America, 765 p.

Fairchild, H. L., 1914, Review of the early history of the Society: Geological Society of America Bulletin, v. 25, p. 17–24. Good history by the Society's second Secretary, presented on GSA's 25th anniversary.

——1932, The Geological Society of America, 1888–1930: A chapter in earth-science history: Geological Society of America, 232 p.

Frondel, Clifford, 1962, The system of numerology (7th edition, Volume III): New York, John Wiley and Sons.

Geological Society of America, 1974, Council rules, policies, and procedures: Geological Society of America, 116 p. Distributed to all members at times of publication; copies available (as of 1980) from GSA headquarters.

Gilluly, James, 1949, Report of the President: Geological Society of America Proceedings for 1948 (1949).

Goldstein, August, Jr., 1973, Investment principles and practices of the Geological Society of America: The Geologist, supplement to v. 8, no. 2, p. 1–9 (May 1973).

Hall, James, 1889, Opening address by the President at the first semiannual meeting held at Toronto, Canada, August 28 and 29, 1889: Geological Society of America Bulletin, v. 1, p. 15–18. Good account of the founding of GSA's predecessor, the Association of American Geologists in 1840.

Heroy, W. B., Jr., and Goldstein, August, Jr., 1976, A brief look at the investment policy of the Geological Society of America: Geological Society of America Geology, v. 4, p. 537. Updates paper by Goldstein (1973).

Holmes, M. E., 1887, The morphology of the carinae upon the septa of the rugose corals [Ph.D. thesis]: University of Michigan, 31 p.

Jefferson, Thomas, 1799, A memoir on the discovery of certain bones of a quadruped of the clawed kind in the western parts of Virginia: American Philosophical Society Transactions, v. 4, p. 246–260. Vertebrate paleontology from a sitting Vice-President of the United States.

Kerr, P. F., 1957, Charles Peter Berkey, 1867–1955: National Academy of Science Biographic Memoirs, v. 30, p. 41.

——1959, Henry R. Aldrich, a master craftsman of the modus operandi: Geotimes, v. 4, no. 5, p. 14–16.

Knopf, E. B., 1946, Memorial of Florence Bascom: American Mineralogist, v. 31, p. 168–172.

Knowlton, F. H., 1913, Memoir of W J McGee: Geological Society of America Bulletin, v. 24, p. 18–28.

LeMoreaux, P. E., and Barksdale, H. C., 1977, Memorial to John Manning Birdsall, 1902–1975: Geological Society of America Memorials, v. 7, 3 p.

Leopold, L. B., 1973, River channel change with time: An example: Geological Society of America Bulletin, v. 84, p. 1845.

Maugh, T. H., II, 1974, Poster sessions: A new look at scientific meetings: Science, 28 June 1974, p. 1361.

Michigan Alumnus, 1899, Class notes: v. 5, no. 9.

Moore, R. C., 1952, Treatise on invertebrate paleontology (report submitted to the Council of the Paleontological Society, 1951): Geological Society of America, 42 p.

Ogilvie, I. H., 1945, Florence Bascom, 1862–1945: Science, v. 102, no. 2648, p. 320–321.

Paige, Sidney, editor, 1950, Application of geology to engineering practice: Berkey Volume: Geological Society of America, 327 p.

Palache, Charles, Berman, Harry, and Frondel, Clifford, 1944, The system of mineralogy (7th edition, Volume I): New York, John Wiley and Sons.

——1951, The system of mineralogy (7th edition, Volume II): New York, John Wiley and Sons.

Penrose, R.A.F., Jr., Memoirs: Typescript, 43 p., single copy in files of Geological Society of America. Covers Penrose's early life, from birth to 1889 (Arkansas Geological Survey).

——1928, In re bust of Professor N. S. Shaler: Unpublished 3-page memorandum dated August 20, 1928, in files of Geological Society of America.

Savage, J. L., and Rhoades, Roger, 1950, Charles Peter Berkey: Application of geology to engineering practice: Berkey Volume: Geological Society of America, p. xi.

Schwarzer, T. F., and Crawford, M. L., 1977, American women in geology (symposium): Geological Society of America Geology, v. 5, no. 8, p. 493–504.

Severson, R. C., Gough, L. P., McNeal, J. M., and Ropes, L. H., 1979, Poster sessions: An alternative to formal oral presentations: Geological Society of America GSA News & Information, v. 1, no. 2, p. 17–18.

Stanley-Brown, Joseph, 1914, History of the Bulletin: Geological Society of America Bulletin, v. 25, p. 24–27. GSA's early publication history by the man who guided it for longer than any other Editor.

Stevenson, J. J., 1914, Events leading up to the organization of the Geological Society of America: Geological Society of America Bulletin, v. 25, p. 15–17. Good history by GSA's first Secretary, presented on the Society's 25th anniversary.

Vaughan, T. W., 1944, Memorial to Robert Thomas Hill [1858–1941]: Geological Society of America Proceedings, 1943, p. 141–168.

Who Was Who in America, 1943, Volume 1 (1897–1942): Chicago, A. N. Marquis Co.

Winchell, Alexander, 1889, Historical sketch of the organization of the Geological Society of America: Geological Society of America Bulletin, v. 1, p. 1–6.

Winchell, N. H., 1914, Review of the formation of geological societies in the United States: Geological Society of America Bulletin, v. 25, p. 27–30. By one of the leaders in the founding of GSA. Presented on the Society's 25th anniversary.

Winchell, N. H., and Hitchcock, C. H., 1888, The proposed geological society: American Geologist, v. 1, p. 395.

Typeset by WESType Publishing Services, Inc., Boulder, Colorado
Printed in U.S.A. by Malloy Lithographing, Inc., Ann Arbor, Michigan